Climate Policy after the 2015 Paris Climate Conference

The 2015 Paris Agreement marked a turning point in the global community's response to climate change. For the first time, almost all the world's nations put forward specific pledges to cut their greenhouse gas emissions with the aim of limiting global warming to 2°C, and ideally to 1.5°C.

The ten contributions in *Climate Policy after the 2015 Paris Climate Conference* provide a powerful and scholarly analysis of how this historic achievement came about. Written by longstanding experts in the field, the chapters explore the merits and flaws of the treaty and the implementation challenges that still lie ahead, including inconsistency between the Paris Agreement's global temperature goals and national pledges; weak provisions on support for developing countries; differing legal interpretations on climate liability; and the potential for trade disputes between more and less ambitious countries.

Altogether, these authoritative contributions provide essential reading for anyone wishing to gain a better understanding of the Paris Agreement, its political origins, and the prospects for international climate policy going forward.

Apart from the addition of an insightful new introduction tracking and analysing developments since 2015, the chapters in this book were originally published as a special issue of the journal *Climate Policy*.

Joanna Depledge is Fellow at the Centre for Environment, Energy and Natural Resource Governance (CEENRG) at the University of Cambridge, and a member of the research network Climate Strategies.

Jorge E. Viñuales is Visiting Professor at LUISS, Guido Carli, and the Harold Samuel Chair of Law and Environmental Policy at Cambridge.

Emma Lees is Professor of Environmental and Property Law, and Wilson Fellow and Dean at Fitzwilliam College, University of Cambridge. She is also the Co-Director of the Cambridge Centre for Property Law and a Fellow of the Cambridge Centre for Environment, Energy and Natural Resource Governance (CEENRG).

David M. Reiner is University Senior Lecturer in Technology Policy at the Judge Business School and Assistant Director of the Energy Policy Research Group (EPRG), University of Cambridge. He is also a Research Associate of the Joint Program on the Science and Policy of Global Change at the Massachusetts Institute of Technology (MIT).

Climate Policy after the 2015 Paris Climate Conference

Edited by
Joanna Depledge, Jorge E. Viñuales, Emma Lees and David M. Reiner

Routledge
Taylor & Francis Group
LONDON AND NEW YORK

First published 2022
by Routledge
2 Park Square, Milton Park, Abingdon, Oxon OX14 4RN

and by Routledge
605 Third Avenue, New York, NY 10158

Routledge is an imprint of the Taylor & Francis Group, an informa business

British Library Cataloguing in Publication Data
A catalogue record for this book is available from the British Library

ISBN: 978-1-032-04351-7 (hbk)
ISBN: 978-1-032-04352-4 (pbk)
ISBN: 978-1-003-19158-2 (ebk)

DOI: 10.4324/9781003191582

Typeset in Minion Pro
by Newgen Publishing UK

Publisher's Note
The publisher accepts responsibility for any inconsistencies that may have arisen during the conversion of this book from journal articles to book chapters, namely the inclusion of journal terminology.

Contents

Citation Information

The following chapters were originally published in the journal, *Climate Policy*, volume 17, issue 1 (2016). When citing this material, please use the original page numbering for each article, as follows:

Chapter 1

Climate policy after the Paris 2015 climate conference
Jorge E. Viñuales, Joanna Depledge, David M. Reiner and Emma Lees
Climate Policy, volume 17, issue 1 (2016), pp. 1–8

Chapter 2

Climate change after Paris: from turning point to transformation
Richard Kinley
Climate Policy, volume 17, issue 1 (2016), pp. 9–15

Chapter 3

The Paris Agreement: resolving the inconsistency between global goals and national contributions
Niklas Höhne, Takeshi Kuramochi, Carsten Warnecke, Frauke Röser, Hanna Fekete, Markus Hagemann, Thomas Day, Ritika Tewari, Marie Kurdziel, Sebastian Sterl and Sofia Gonzales
Climate Policy, volume 17, issue 1 (2016), pp. 16–32

Chapter 4

Precaution and post-caution in the Paris Agreement: adaptation, loss and damage and finance
Anju Sharma
Climate Policy, volume 17, issue 1 (2016), pp. 33–47

Chapter 5

The Paris Agreement: China's 'New Normal' role in international climate negotiations
Isabel Hilton and Oliver Kerr
Climate Policy, volume 17, issue 1 (2016), pp. 48–58

Chapter 6

Responsibility and liability for climate loss and damage after Paris
Emma Lees
Climate Policy, volume 17, issue 1 (2016), pp. 59–70

Chapter 7

Small group, big impact: how AILAC helped shape the Paris Agreement
Guy Edwards, Isabel Cavelier Adarve, María Camila Bustos and J. Timmons Roberts
Climate Policy, volume 17, issue 1 (2016), pp. 71–85

Chapter 8

US-proofing the Paris Climate Agreement
Luke Kemp
Climate Policy, volume 17, issue 1 (2016), pp. 86–101

Chapter 9

Global trade and promotion of cleantech industry: a post-Paris agenda
John A. Mathews
Climate Policy, volume 17, issue 1 (2016), pp. 102–110

For any permission-related enquiries please visit:
www.tandfonline.com/page/help/permissions

Notes on Contributors

The following affiliations are from the time the journal articles were originally published.

Isabel Cavelier Adarve, Permanent Mission of Colombia to the United Nations, 140 East 57th Street, New York, NY 10022, USA.

Maria Camila Bustos, Institute at Brown for Environment and Society, Brown University, 85 Waterman Street, Providence, RI 02912, USA.

Thomas Day, NewClimate Institute, Am Hof 20-26, Cologne 50667, Germany.

Joanna Depledge, Centre for Environment, Energy and Natural Resource Governance, University of Cambridge, Cambridge, UK.

Guy Edwards, Institute at Brown for Environment and Society, Brown University, 85 Waterman Street, Providence, RI 02912, USA.

Hanna Fekete, NewClimate Institute, Am Hof 20-26, Cologne 50667, Germany.

Sofia Gonzales, NewClimate Institute, Am Hof 20-26, Cologne 50667, Germany.

Markus Hagemann, NewClimate Institute, Am Hof 20-26, Cologne 50667, Germany; Faculteit Geowetenschappen, Utrecht University, Heidelberglaan 2, 3584 CS Utrecht, The Netherlands.

Isabel Hilton, Chinadialogue, Suite 306, Grayston Centre, 28 Charles Square, London N1 6HT, UK.

Niklas Höhne, NewClimate Institute, Am Hof 20-26, Cologne 50667, Germany; Environmental Systems Analysis Group, Wageningen University, PO Box 47, Wageningen 6700AA, The Netherlands.

Luke Kemp, Fenner School of Environment and Society, Australian National University (ANU), Building 141 Linnaeus Way, Canberra, ACT, 2601, Australia.

Oliver Kerr, UCL Energy Institute, The Bartlett, UCL Faculty of the Built Environment, Central House, 14 Upper Woburn Place, London WC1H 0NN, UK.

Richard Kinley, UNFCCC Secretariat, UN Campus, Platz der Vereinten Nationen 1, 53113 Bonn, Germany.

Takeshi Kuramochi, NewClimate Institute, Am Hof 20-26, Cologne 50667, Germany.

Marie Kurdziel, NewClimate Institute, Am Hof 20-26, Cologne 50667, Germany.

Emma Lees, Department of Land Economy, University of Cambridge, Cambridge, UK.

John A. Mathews, Macquarie Graduate School of Management, Macquarie University, Sydney, NSW, Australia.

David M. Reiner, Energy Policy Research Group, Judge Business School, University of Cambridge, Cambridge, UK.

J. Timmons Roberts, Institute at Brown for Environment and Society, Brown University, 85 Waterman Street, Providence, RI 02912, USA.

Frauke Röser, NewClimate Institute, Am Hof 20-26, Cologne 50667, Germany.

Anju Sharma, Oxford Climate Policy, 139 Cumnor Hill, Oxford OX2 9JA, UK.

Sebastian Sterl, NewClimate Institute, Am Hof 20-26, Cologne 50667, Germany.

Ritika Tewari, NewClimate Institute, Am Hof 20-26, Cologne 50667, Germany.

Jorge E. Viñuales, Department of Law, LUISS, Rome, Italy; University of Cambridge, Cambridge, UK.

Carsten Warnecke, NewClimate Institute, Am Hof 20-26, Cologne 50667, Germany.

Introduction: international climate policy after Paris – an update on a changing world

Joanna Depledge

Introduction: a changing world

The international climate policy landscape has evolved significantly since the Paris Agreement was adopted to great fanfare at the 21st Conference of the Parties (COP 21) to the United Nations Framework Convention on Climate Change (UNFCCC) on 12 December 2015. The overall direction of travel, however, is mixed. The latest scientific assessment from the Intergovernmental Panel on Climate Change (IPCC, 2021) has confirmed that human-induced climate change is intensifying, and is exacerbating extreme weather events, such as heatwaves and floods. Global emissions have continued to rise, albeit more slowly than in the decade to 2010. While the Paris Agreement entered into force in record time and its rulebook was largely agreed two years later, the effectiveness of its "ratcheting up" mechanism – intended to spur an upward spiral of rising ambition – remains in doubt. At the same time, political bickering at subsequent conferences – both in person and online – point to an evaporation of the constructive "spirit of Paris". Although the (temporary) US withdrawal from the treaty did not trigger further defections, the absence of an engaged US Presidency has taken its toll. A wave of "climate emergency" and "net zero" declarations has spread among countries, cities and organisations, yet the implementation of such good intentions remains to be tested. Outside the world of climate diplomacy, the rise of the youth movement inspired by teenager Greta Thunberg has boosted public awareness and support for strong climate action like never before, but once again, this has yet to translate into concrete, short-term policy decisions. The cost of solar photovoltaics (PV) and wind energy has tumbled, yet coal mines are still being opened and oil exploration continues, although at a reduced rate. Most dramatically, since 2020, the Covid-19 pandemic has turned all policy assumptions upside down. The climate world is truly a different and more complex place than it was in 2015.

The nine chapters of this book were written in 2016, in the immediate afterglow of the Paris Conference. They sought to analyse the Paris Agreement from multiple angles – the diplomatic process and politics behind the treaty's successful negotiation, its strengths and weaknesses, unresolved questions and ambiguities, and prospects for its future. The chapters have stood the test of time, pointing to many issues that remain, or have since become, salient. Even if some questions raised never came to pass, or have dropped off the political agenda, it is still instructive to review developments since the Paris Agreement's adoption in light of the assessments, interpretations, hopes and fears that were current at the time. Against this background, this introduction explores the different ways in which international climate policy has evolved since 2015, based on key themes discussed in the other nine chapters.

The global goals: embracing and tightening

A key achievement of the Paris Agreement was to establish more precise and ambitious global mitigation goals. Countries agreed to limit global temperature rise to "well below" 2 °C above pre-industrial levels, and to "pursue efforts" towards 1.5 °C (Article 2.1(a)). The 2 °C threshold had already been agreed in the 2010 Cancun Agreements (building on the 2009 Copenhagen Accord). However, its inclusion in a legally-binding international treaty, the qualifier "well below", and the addition of 1.5 °C, albeit in aspirational terms, "ask[ed] for mitigation ambition to be raised to a whole new level" (Höehne et al., 2017 and this volume). These temperature goals were further reinforced by the aim "to reach global peaking" of emissions "as soon as possible" and "to achieve a balance" between emissions and removals (in effect, net zero emissions) "in the second half of the century" (Article 4.1).

Since their adoption, these global goals – especially the temperature limits and net zero – have become increasingly embedded in public discourse on climate change, beyond the immediate diplomatic arena. The term net zero, in particular, seems to have caught the public imagination, proving, as Kinley (2017 and this volume) predicted, to be "more popularly understandable". At the time of writing, around 60% of global emissions were covered by announced or planned net zero targets, most for 2050, backed up by legislation (or legislation in process) in around a dozen countries (including the EU-27). In addition, a database maintained by the UN Climate Secretariat recorded net zero target announcements from hundreds of cities, organisations, and more than a thousand private sector businesses across the world. This rallying around net zero has been a major step forward since Paris. However, definitions are often fuzzy, raising doubts as to the true levels of ambition, while such declarations are rarely accompanied by commensurate policies.

The temperature goals have also become embedded, and in effect toughened, since Paris. When Parties settled on a "well below" 2°C goal in Paris, with 1.5°C framed in more aspirational terms, the IPCC was invited to prepare a special report "on the impacts of 1.5°C" (decision 1/CP.21, para. 21). This responded to the concerns of least developed countries (LDCs) and small island developing states (SIDS) in particular, who feared that even 1.5°C of warming would pose a significant threat. The resulting IPCC special report, published in 2018 (IPCC, 2018), was generally well-received across the international community. Representing an unprecedented collaboration across the IPCC's three working groups (physical science, impacts, and mitigation), it established, in stark terms, that limiting warming to 1.5°C would considerably reduce the risks to vulnerable communities and ecosystems relative to the more severe impacts associated with 2°C. Although Saudi Arabia and the US sought to weaken its findings before adoption, and to downplay its importance at the subsequent Katowice Climate Conference (COP 24) (BBC, 2018), the IPCC 1.5°C special report has proved highly influential. In effect, it has prompted a shifting of the temperature goalposts towards the lower end of the thermometer. Businesses and environmental groups are increasingly referring to a 1.5°C, rather than well below 2°C goal,[1] while climate change legislation or policy modelling is being revised in many countries to take account of a stricter 1.5°C limit. Christiana Figueres, head of the UN Climate Secretariat during the Paris Agreement negotiations, declared in April 2021 that "1.5 is the new two degrees",[2] while a key goal of the UK Presidency of the Glasgow Climate Conference (COP 26) has been to "keep 1.5 degrees within reach".[3] Kinley has been proven right, when he suggested, "aspirations create expectations and are thus extremely important in politics" (2017 and this volume).

The nationally determined contributions (NDCs): an upward spiral of ambition?

The Paris Agreement's nationally determined contributions (NDCs) reflect a well-known trade-off in international treaties, between scope and strength of commitments. The NDCs are universal in scope – all Parties, irrespective of development status, must table an NDC at regular intervals – which in itself is a great achievement. However, this universality came at the expense of weakness of commitments: in legal force (no obligation to meet the pledges), comparability (absence of all but the most general common metrics), and ambition. And this latter weakness constitutes the central predicament of the Paris Agreement: the collective level of ambition of the declared NDCs is inconsistent with the global goals, as explained by Höehne et al. (2017 and this volume), and by many others since (e.g. UNEP, 2020). The NDCs, even if fully implemented (which is uncertain), add up to "well *above*", not "well below", a 2°C world. However, as Höehne et al. (2017 and this volume) and others have argued, the promise, and premise, of the Paris Agreement is that an upward spiral of rising ambition will be triggered by its timetable of rolling NDC revisions, combined with rigorous reporting and the five yearly "global stocktake" of collective impact that launches in 2023. The upward spiral is expected to be driven principally by interaction between the forces of "naming and shaming", leading by example, civil society demands and low-carbon market trends (Viñuales et al., 2017 and this volume), while the very process of preparing an NDC should build capacity within nations to formulate more effective climate policy (Höehne et al., 2017 and this volume).

It is too early to tell if this promise will be realised. The timetable for revising the NDCs was thrown into disarray by the Covid-19 pandemic, which reduced government capacity to focus on climate change pledges, and resulted in the postponement of COP 26, initially scheduled to be held in Glasgow in November 2020. Although decision 1/CP.21 (para. 23 and 24) states that NDCs should be updated or revised "by 2020", this was widely interpreted as meaning "by COP 26". The result is that, as of July 2021, less than 100 new or revised NDCs had been submitted (including from the EU-27), of which less than half represented a clear uptick in ambition (with the US being the most notable). However, several large emitters, including China and Japan, were expected to table revised NDCs in 2021 based on net zero targets. If all the declared net zero targets were translated into NDCs, it would bring the world within striking distance of the 2°C threshold (Climate Action Tracker, 2020). Despite these promising signals from some, many other large emitters are proving impervious so far to pressure from peers or civil society. Furthermore, the real impact on the climate will come from the implementation of targets and NDCs, not their declaration.

Adaptation, and support to developing countries: slow, incremental progress

The NDCs are only one component of the Paris Agreement, and of the wider international effort to address climate change. In her contribution to this volume, Sharma draws attention to the important issue of climate finance, providing a critical assessment of the Paris Agreement's meagre provisions, which in effect merely reiterate existing commitments. Public financial support for developing countries is a very longstanding, fractious topic in the climate change regime – indeed in most corners of the international landscape – and little has changed since the Paris Agreement's adoption. Climate finance is on an upwards trend, but the target for developed countries to provide $100 billion a year by 2020 to support developing countries – which was first mooted in 2009, agreed in 2010, and reiterated in the Paris Agreement decision text – has likely been missed (OECD, 2021). Disputes still rage about the share of loans versus grants, additionality relative to development aid, and concerns over opacity and double-counting in reporting. As Höehne et al. (2017 and this volume) warn, the NDCs of many smaller developing countries are conditional upon the receipt of sufficient financial (and other) assistance, and a shortfall could therefore threaten NDC implementation and the raising of ambition.

The Paris Agreement includes a dedicated article and new global goal on adaptation (Article 7 and 2.1(b)). This was seen by some as a major step towards placing adaptation on a par with mitigation (e.g. Kinley et al., 2017 and this volume), although others, notably Sharma (also 2017 and this volume), regretted the continued focus on adaptation planning and reporting (rather than implementation), and repetition of "generalities" on adaptation finance (rather than setting a quantified goal). There has been some progress since 2015, including confirmation that the Paris Agreement can use the Adaptation Fund originally established under the Kyoto Protocol, and adoption of guidelines to prepare "adaptation communications". However, only limited headway has been made towards answering the Paris Agreement's call for "scaled up financial resources" to achieve a "balance" (Article 9.4) between adaptation and mitigation. Despite a signficant rise in 2019, funding for adaptation still accounts for only about a third of total climate finance (OECD, 2021). Overall, there is potential for the Paris Agreement to bring a real uplift to adaptation, but there is still a long way to go before what Kinley et al. (2020) call "the historic gap in funding for adaptation measures" is filled.

The Paris Agreement marked a pivot towards greater recognition of the potentially devastating impacts of climate change on vulnerable communities through its provisions on "loss and damage". Famously, these include both a dedicated article on the topic, and a rider in decision 1/CP.21, qualifying that this Article 8 "does not involve or provide a basis for any liability or compensation" (paragraph 51). These elements of the Paris deal have provided considerable fodder for subsequent legal analysis, with varying interpretations of their import and potential for influencing policy (e.g. Broberg and Martinez Romera, 2020). Lees provides one such assessment in this volume, arguing that it is possible to build upon Article 8 in constructive ways that would bypass the strictures of paragraph 51. These provisions, and their varying interpretations, remain untested so far.

Cases of climate litigation, however, have multiplied since the adoption of the Paris Agreement, with over 1,000 filed since 2015; although these are still dominated by litigation in the US, cases elsewhere, including in the Global South, are rising (Setzer and Higham, 2021). Some of these, including the landmark *Urgenda* case in the Netherlands that was upheld in 2018 and the ongoing *People's Climate Case* against the European Union, have invoked the Paris Agreement's 2°C or 1.5°C global goals, and failure of their jurisdiction to take sufficient action to meet their fair share of it. This trend towards climate litigation, increasingly on the basis of violation of human rights (UNEP/SCCL, 2020), holds great potential for ensuring the implementation of national action, even in the absence of legally-binding NDC commitments or strong compliance provisions. Whether such legal action ever takes the form of liability or compensation claims for loss and damage remains to be seen.

And so, similar to climate finance and adaptation, there have been advances since Paris on loss and damage, but they are incremental rather than transformational. Information gathering and exchange continues under the 2013 Warsaw Mechanism, but the fundamental question of how to secure adequate funding to address the unavoidable and devastating impacts of climate change remains unanswered, as it has for almost 30 years.

The "rule book" and Article 6: exposing unresolved conflicts

The first diplomatic task after entry into force of the Paris Agreement was to negotiate and decide upon the rulebook to implement its provisions, in particular on transparency (reporting and review) and the Article 6 mechanisms for "voluntary cooperation" (in short, elaboration of rules for international emissions trading or project-based emission crediting, and a rather hazily defined "sustainable development mechanism"). The negotiations on almost all elements of the rulebook were adopted at the 2018 Katowice Conference, albeit not without intensive and often acrimonious debate; rules for donor countries to report on expected financial flows, for example, were subject to particularly tense exchanges. Such acrimony confirmed warnings voiced by Viñuales et al. (2017 and this volume) that the unity shown in Paris had been largely achieved not by resolution of deeply-held differences, but rather by papering over those rifts with the diplomatic tools of clever wording and deferral of difficult issues (e.g. see Biniaz, 2016).

Viñuales et al.'s warnings were thrown into focus by events at the 2019 Chile/Madrid Conference, where delegates once again failed to finalise the outstanding elements of the Katowice rulebook, notably on rules for the Article 6 mechanisms and the definition (or not) of common timeframes for future NDCs. This was despite the negotiations going into 40 hours of overtime and thereby lasting longer than any other previous climate conference (never known for their prompt conclusion). The Article 6 mechanisms, and other outstanding elements of the Paris rulebook, must now be finalised at the Glasgow Climate Conference. However, the hiatus in formal negotiations resulting from the Covid-19 pandemic has done little to prepare the ground for a more successful outcome.

Political realignments: the rise of the vanguard?

Some of the constructive political advances and shifts in alignments that were cemented in Paris, however, have stuck. One of the iconic images of the Paris Conference is of a large group of delegates, led by senior negotiators from the Marshall Islands and (somewhat improbably) the US, entering the plenary hall together, as the self-styled "high ambition coalition" (HAC). Support from the HAC, at its peak including over 100 delegations from the EU, LDCs, SIDS, African Group, and Independent Association of Latin American and the Caribbean (AILAC), was credited with securing agreement on the 1.5°C aspirational target, among other demands of vulnerable countries. In their contribution, Edwards et al. (2017 and this volume) draw attention to the importance of progressive smaller countries (such as those in AILAC, the focus of their study) in enabling such bridge-building "coalitions of ambition" – importantly spanning both developed and developing countries. This realignment of political forces, including the growing influence and assertiveness of smaller countries, began before Paris, and has continued since. The new political fault lines are no longer (at least for most issues) between developed and developing countries, but between the willing and the un – or less – willing, or the "vanguard vs the laggards" (Climate Change News, 2019). New and sometimes surprising alliances are emerging. At the Chile/Madrid Conference, for example, a group of 31 countries, led by Costa Rica, released the so-called San José principles aimed at securing environmental integrity in any deal on Article 6 mechanisms. This self-styled "unconventional group" cut across traditional coalitions, including countries from the EU, Latin America, and SIDS, plus New Zealand.[4] In another political phenomenon, "vanguard" countries are increasingly refusing to consent to weak outcomes; where once it was the "laggards" who would block decisions until they were watered down, "vanguard" countries, including small ones acting in coalitions, are now blocking vacuous texts until they are strengthened. At the Katowice Conference, for example, a large group of countries spanning regional and political divides refused to bow to pressure to adopt a weakened decision that failed to "welcome" the IPCC report on 1.5°C (Climate Change News, 2018). This new-found resolve to expose, rather than paper over, political differences – especially when these pit the vast majority of more willing nations against a handful of laggards – may be painful, but it surely represents an important step to finally making concrete progress in the international climate regime, rather than settling for elegant wording and constructive ambiguity, which achieve little more than delay.

China: the "new normal" continues

One of the main political forces that secured the consensus adoption of the Paris Agreement, and shaped its contents, was the alliance that was forged between the US and China – at head-of-state level – already prior to the Paris Conference (Hilton and Kerr, and Kinley, both 2017 and this volume). In many ways, the interests and diplomatic stance of the two highest emitters were broadly aligned in Paris, contrasting with the situation at the 2009 Copenhagen Conference. Neither country wanted to take on legally-binding, quantified targets in Paris, but both were under great pressure to assume a much stronger role in bending the global emissions curve, in line with their high share of emissions and economic might. Hilton and Kerr (2017 and this volume) explain how, by 2015, cutting domestic carbon emissions was now aligned with China's "new normal", pivoting away from energy-intensive industry and exports as an engine of growth, and instead turning to the service sector, domestic consumption, and high-tech (including green) innovation. Already by 2015, China was a world leader in solar PV and battery technology, and had launched emission trading pilot schemes in nine jurisdictions. Under President Obama, the US was seeking to rejoin the international climate change regime as a full partner, after its repudiation of the Kyoto Protocol under President George W. Bush, followed by clumsy diplomacy in Copenhagen. The Chinese and US positions in Paris were not completely in step – China never joined the HAC and was more wary of strong monitoring and review provisions – but it is unlikely that the Paris Agreement would ever have been finalised without both countries being committed to striking a deal and along broadly similar lines (Viñuales et al., 2017 and this volume).

China's "new normal" domestic commitment to cleaner economic development has largely persisted since Paris, albeit peppered with contradictions and mixed messages. China has met its 2020 pledge under the 2010 Cancun Agreements to reduce the carbon intensity of its economy 40–45% from 2005 levels, and to increase the share of non-fossil fuels in primary energy consumption to about 15%. It is also likely to overachieve on its 2030 NDC pledge, peaking its emissions before that year. Its nationwide emissions trading scheme (ETS), while delayed and scaled back, will overtake the EU ETS as the largest such scheme in the world when fully operational (probably in 2025). In September 2020, China took the momentous step of pledging to achieve "carbon neutrality" by 2060, indicating a clear direction of travel towards progressively stronger targets, in line with the Paris Agreement's approach. Plateauing emissions around the time of the Paris Conference raised hopes that Chinese emissions may have peaked, but these were dashed as the Chinese emissions curve resumed its upwards trend from 2017, with a particularly steep rise of over 3% in 2019, due mostly to increased coal consumption (Olivier and Peters, 2020). As of 2020, China had the rather improbable distinction of being the world's largest funder and builder of both new renewable energy and new fossil fuel capacity.[5]

On the international political stage, China remains an ambivalent partner. Despite ramping up its domestic target, China continues to hesitate when it comes to strengthening the global regime, or making collective declarations of global ambition. China, for example, was named as one of "the main blockers" of the Chile/Madrid Conference by the Chilean COP President (Climate Change News, 2020). Just a few months later, however, President Xi Jinping signalled a new approach when he announced China's 2060 carbon neutrality goal to the UN General Assembly, before the US Presidential election. The timing and audience of the surprise announcement sent out a clear message: China is prepared to forge ahead with its domestic decarbonisation, irrespective of the US stance. While partnership with the US was critical in shaping China's position in Paris, this no longer holds true.

The US: dramatic policy reversals

After the many diplomatic successes of 2015 and 2016, multilateralism was dealt a heavy blow by the election of "America first" Donald Trump as US President in November 2016. Less than a year after President Obama formally accepted the Paris Agreement by executive order in a coordinated move with China, newly elected President Trump announced that the US would withdraw from the treaty at the earliest opportunity. For veteran observers of the climate change negotiations, this was déjà vu. In 2001, newly elected President George W. Bush repudiated the Kyoto Protocol that had been adopted under his predecessor Bill Clinton. As with the Paris Agreement, the US had participated very actively in the negotiations on the Kyoto Protocol, so that the treaty broadly mirrored US preferences, not least with the inclusion of market mechanisms. Yet the Clinton Administration never sent the Kyoto Protocol for ratification to the Senate, where it faced certain rejection, and soon after President George W. Bush's election, his administration said it had "no interest" in implementing the treaty (*The Guardian*, 2001). The Paris Agreement was similarly designed largely according to US negotiating demands, with "bringing the US back in" a fundamental – if formally unspoken – aim. In particular, as Kemp explains in this volume, it was specifically drafted so that the US President could formally accept it by executive order – literally by the stroke of a Presidential pen – bypassing the need for Senate ratification. Two treaties designed to enable US participation, two treaties rejected by a subsequent (Republican) President. This eventuality was foreseen by Kemp, who prophetically warned that "any future Republican president is likely to repeal such an agreement, effectively making the US a non-party". Viñuales et al. and Sharma (2017 and this volume) both also cautioned against the expediency of presidential executive orders, with Sharma noting that "when ratification is sought without ensuring broad political support, problems in implementation and future withdrawal, should the opposition come to power, are a distinct possibility". By an accident of timing, US withdrawal from the Paris Agreement took formal effect just a few days before the election of President Joe Biden. As one of his first acts in office, President Biden signed yet another executive order bringing the US back into the Paris Agreement.

The effect is that, for four years, international efforts on climate change had to contend with a climate-sceptic US administration that was nominally still Party to the Paris Agreement, but in practice hostile to it. Domestically, the Trump administration sought to undo existing climate and environmental legislation, while boosting fossil fuel industries, notably coal (in the end with limited success (Galik, deCarolis and Fell, 2017; Mendelevitch, Hauenstein and Holz, 2019)). At the international level, fears that announced US withdrawal would lead to a rash of defections from other climate laggards proved unfounded. Although Russia, the fourth largest emitter, was in no rush to ratify, it eventually did so, and the Paris Agreement now enjoys almost universal participation.[6] The Paris Agreement's weaknesses here may indeed have been its strengths. Because, in practical terms, it does little more than to encourage and codify domestic action, the costs of joining are low, and there is less reason to defect. At the same time, the absence of a cap-and-trade system binding countries together (as under the Kyoto Protocol) meant that the US withdrawal had no immediate impact on the efforts of others. Concerns that the US delegation might become an antagonistic blocking force in the negotiations pending its formal withdrawal were also not borne out. The US delegation, which retained many officials from previous administrations, mostly kept a low profile, continuing to pursue well-known US positions (e.g. in support of transparency, flexibility, and common obligations for all countries). China's net zero announcement – deliberately distancing itself from the US and the coordinated action that had characterised the period 2014–2016 – is a clear indication that the US is not indispensable, at least as a political partner.

However, although the climate change regime has proved resilient to US withdrawal, it has not emerged unscathed. The four-year absence of the world's second largest economy and emitter has unquestionably led to a slackening of the pace in the climate change regime. Delays to the submission of updated NDCs, the weakness of many of those pledges, inability to resolve what should be relatively minor issues, and a general cooling of the negotiating atmosphere, cannot solely be laid at the door of a disengaged US. But it will certainly not have helped. The hostility of the US provided political cover for the more reluctant nations to take their foot off the pedal, or to raise objections when they might otherwise have joined the consensus. After all, the developed countries are required to "take the lead" in cutting emissions under the UNFCCC (Articles 3.1 and 4.2(a)) and "should continue" doing so under the Paris Agreement (Article 4.4), and if the main emitter is not fulfilling this side of the bargain, then the carefully balanced political deal no longer holds. At the same time, there will have been none of the diplomatic outreach from the US that is so important to building and sustaining a global consensus. In practical terms, the Trump administration immediately cancelled all financial contributions to the Green Climate Fund leading to reduced support for climate action in developing countries.[7]

The US has now rejoined the Paris Agreement, and President Biden has laid out a suite of ambitious climate policies, domestic and international. This is of course welcome, but there is reason to be cautious. "President Biden has not, for example, rushed to pay the outstanding US debt to the Green Climate Fund, to the great disappointment of developing countries and with potentially serious implications for trust in the administration". Veteran climate observers will also be reminded of the first climate conference in 2009 following the election of President Barack Obama, when US head of delegation Todd Stern announced to a cheering plenary that "we are very glad to be back, we want to make up for lost time" (Grist, 2009). This lasted only eight years. Because the Paris Agreement can be accepted and rejected by the prevailing US administration by Presidential executive order, it is highly vulnerable to future changes in Presidential views. The same is true for many forms of environmental legislation at the domestic level, while legislation requiring Congressional approval still faces the challenge of deeply partisan attitudes to climate change between Republicans and Democrats. What if the next President is a climate sceptic, who yet again withdraws from the Paris Agreement? The Paris Agreement was resilient to one term of a hostile US President, but could it survive prolonged opposition from the world's second largest emitter, and key economic and political player on the global stage? The US has proved to be an unreliable partner in the international climate change regime, and its authority and any leadership claims have been badly damaged. Given US exceptionalism, the question also arises as to how much the other 190+ parties should bow to US demands in the negotiations, in the hope of ensuring its participation. The Paris Agreement itself would have looked quite different, without the need to accommodate the US.

Trade measures: on the horizon at last?

Such questions led Kemp (2017 and this volume) to argue for strategies to "US-proof" the Paris Agreement, or indeed "proof" it against any powerful defectors. These include the use of trade measures, such as border carbon adjustments, against non-Parties to the treaty. Such trade measures would have as their purpose both to encourage non-Parties to join the Agreement, or at least to adopt commensurate climate policies, and to protect countries with strong emission regulation in place from competition from unregulated regions. The option of introducing such trade measures, and in particular whether they would be compatible with World Trade Organisation (WTO) rules, is taboo in the climate change negotiations themselves, but has long been debated in academic circles (e.g. Monjon and Quirion, 2011). Mathews (2017 and this volume) continues the conversation from a different angle, calling for integration between the WTO and climate change regime, in order to forestall potential trade disputes over domestic promotion and protection of low-carbon industries. This broad debate over trade has gained in salience since Paris, with the EU announcing firm plans to introduce a Carbon Border Adjustment Mechanism as part of its Green New Deal. US President Biden proposed something similar in the run-up to his election. Trade measures may finally have their day, and be tested in practice, rather than just in academic papers.

Civil society: now ahead of the game?

A second strategy proposed by Kemp to "US-proof" the Paris Agreement is to harness the greater commitment of US civil society. This, again, proved prescient. Contrary to President George W. Bush's repudiation of the Kyoto Protocol in 2001, which at the time was almost universally welcomed by US businesses, President Trump's announcement of withdrawal from the Paris Agreement triggered a wave of declarations of continued commitment to the treaty from a wide range of US states, cities, businesses, and institutions (e.g. Arroyo, 2018). Even ExxonMobil, which had funded a multi-million dollar misinformation campaign against climate change and the Kyoto Protocol 20 years before, expressed support for the Paris Agreement.[8] At climate change conferences between 2017 and 2019, there seemed to be two US delegations – the official, subdued one representing the Trump administration, and the vocal, active, and ambitious one, representing a wide ranging coalition of non-state actors determined to show continued support from large sections of US society.

This greater engagement of civil society extends much further than the US. Harnessing civil society support for stronger governmental action, and showcasing concrete initiatives by businesses, sub-national governments, and industry organisations, was a deliberate strategy of the COP Presidencies and UN Climate Change Secretariat in the run-up to Paris. Various initiatives were launched to register and parade non-state pledges on climate change, in the hope of persuading governments (at least the democratic ones among them) that there was demand from voters for a strong agreement. Negotiation leaders have sought to maintain the momentum since, sponsoring or showcasing an array of action plans and programmes, also involving the wider UN system under the UN Secretary-General. This reflects an important and positive trend since Paris – the increasing take-up of climate action and ambition among stakeholders, often beginning with, or centred around, the declaration of net zero or other emission reduction targets (Kuramochi et al., 2020; Lui et al., 2021). Although precise figures are uncertain and continuously rising, at the time of writing, the Global Climate Action portal on the UN Climate Change Secretariat website registered over 20,000 actors (businesses, cities, universities, investors, other organisations) undertaking nearly 30,000 actions. The next step must be to develop ways of tracking the implementation of such initiatives, to ensure that they are followed through, along with methodologies to better understand their concrete impact on emissions (Hale et al., 2020).

A particularly consequential development since Paris has been the emergence of a more engaged financial sector, including moves to "green" central banks and sector-wide initiatives to encourage more transparency in the holding of fossil fuel liabilities (which could potentially become "stranded"). A Net Zero Alliance of Asset Managers, for example, was launched at the UN Secretary General's Climate Action Summit in 2019, and a UN Special Envoy on Finance and Climate Action has been appointed. The rise of a civil society-led divestment movement, pushing financial institutions (including pension funds) to withdraw funding for fossil fuel projects, is also starting to bite (e.g. see Curran (2020) on the Australian Carmichael coal mine).

Another related trend that began before Paris and has accelerated in momentum since (at least pre-Covid-19), has been the rise of mass social movements and public demonstrations in support of stronger climate policies. Climate marches in the run-up to Paris saw tens of thousands of people take to the streets in organised demonstrations in cities around the world. These were credited with helping to build political pressure for a deal in Paris (Kinley, 2017 and this volume). Popular climate change protests have taken on a new dimension since then, with the emergence of a distinct youth movement inspired by Swedish teenager Greta Thunberg. Thunberg first came to public attention in 2018, when she began to miss school on Fridays, to instead sit in front of the Swedish Parliament to demand more action on climate change. Her lone protest soon snowballed into a "Fridays for Future" movement, involving school students worldwide. Just before the pandemic brought public demonstrations to an abrupt end, thousands of people marched through Madrid in December 2019 during the Chile/Madrid Conference, with their sense of urgency and big-picture demands contrasting with the political timidity and quarrelling over minutiae in the conference halls.

Overall, there is the sense that public awareness of, and support for, rapid decarbonisation has risen sharply since Paris, just as the climate change regime itself is stagnating. As Kinley et al. (2020) observe, "whereas in its early years the UNFCCC process was ahead of public consciousness, it now lags behind". Streck (2020:145) agrees that the process is now "out of step" with public opinion. This has happened before, with the diplomatic UN climate change process alternately leading, then lagging behind, the "real world" (Depledge, 2006). It is when the two are in step in their levels of ambition – as they were in Paris – that the most progress can be made.

The wider international system: getting on board?

An effective global response to climate change requires the entire international community also to be in step, redeploying all the financial, operational, and political tools of the UN system and related intergovernmental institutions towards a low-carbon and climate-resilient future. Perhaps surprisingly, this all-of-UN approach has been slow to materialise (Kinley et al., 2020). Much as individual national governments have found mainstreaming climate change concerns across different departments a challenge (if indeed they have tried at all), the UN system has also struggled to appreciate the all-encompassing nature of the problem and integrate mitigation and adaptation objectives into its programmes. For too long, intergovernmental development banks have continued to fund fossil fuel projects, while UN agencies have failed to seize the opportunity to promote effective implementation of the climate change treaties at country-level (Kinley et al., 2020). This is now changing. The adoption of the UN Sustainable Development Goals (SDGs) just a few months before the Paris Conference, with climate action as goal 13, has embedded a much greater understanding of the links between climate change, development, and human aspirations more broadly, feeding through to a more integrated response among UN agencies. The adoption of the Sendai Framework for Disaster Risk Reduction 2015–2030, at the Third UN World Conference on Disaster Risk Reduction also in 2015, similarly helped to elevate climate change risks within the UN system.

The post-Paris period has seen three long overdue breakthroughs within the wider climate change policy complex, on important sectors that had previously managed to escape international rule making.

Marking another diplomatic triumph for the international community soon after Paris, the Kigali Amendment to the Montreal Protocol on Substances that Deplete the Ozone Layer was adopted in October 2016. The Kigali Amendment addresses hydrofluorocarbons (HFCs), which have been rolled out since 1990 as alternatives to some of the ozone-depleting substances (ODS) being phased out under the Montreal Protocol ozone regime. However, although HFCs are not ozone-depleting, they are still potent greenhouse gases (but less so than the ODS they replace). This important interaction between the world's two treaty systems on the global atmosphere was first raised in the climate change regime in 1998. Finally, after many years of difficult negotiations, the Montreal Protocol parties (which effectively mirror the climate change parties) agreed to phase down HFCs by around 85% by 2050, in favour of non-global warming alternatives. If fully implemented (and the Montreal Protocol has a good track record in this respect), this should lower global temperature by 0.4°C by 2100 (UNEP/Ozone, 2019). This case of one regime acting (albeit belatedly, and some would say still timidly) to mitigate its impact on another is a welcome demonstration that it is possible for different parts of the international system to act coherently.

The other two rule-making advances concern emissions from international aviation and maritime transport, each accounting for about 2% of global emissions, more than the individual national totals of Australia, Germany, or Saudi Arabia. How to address emissions from these sectors has been a vexed question in the climate change negotiations for decades, with any semblance of meaningful discussion quickly shut down by oil exporting countries, with the tacit support of the US. The issue was in effect sub-contracted out by the Kyoto Protocol to the respective UN-based sector bodies, the International Civil Aviation Organisation (ICAO), and the International Maritime Organisation (IMO). Although climate change was taken up and debated by these bodies, and aspirational statements issued, progress was glacial. The former UN Climate Change Executive Secretaries (see Kinley et al. (2020)) assessed "the slow and half-hearted responses" as "shameful", the same hard-hitting descriptor employed by Marshall Island Foreign Minister Tony de Bruin in relation to the IMO in the run up to the Paris Conference.[9] Then, from 2012, the EU brought aviation under the scope of its ETS, and only agreed to exempt international flights to and from the region on the understanding that a global response would be developed. This no doubt focussed the minds of industry and ICAO delegates, and in October 2016, the ICAO Council adopted a resolution establishing a Carbon Offsetting and Reduction Scheme for International Aviation (CORSIA), aimed at achieving "carbon neutral growth" from 2020. Under CORSIA, airlines are required to offset any rise in their CO_2 emissions from 2021 onwards, by, for example, purchasing emission reduction credits on the international market. Reaching agreement on such a scheme was itself a major achievement, involving as it did not only the reconciliation of contrasting government positions, but also actors from the aviation industry; as

Anjaparidze (2019) points out, no other sector has an overall cap on emissions. However, CORSIA as it stands is weak, and its provisions are not commensurate with a 2°C world: participation is voluntary up to 2027, the plan does not tackle non-CO_2 impacts on climate change, and the need to cut absolute emissions is not acknowledged (Larsson et al., 2019). Moreover, the amount of credible offsets required could be astronomical, far outstripping the supply likely to be generated under the climate change regime, even when (if) agreement is reached on rules under Article 6, and even setting aside the concerns over ethics and environmental integrity that continue to dog international carbon markets.

All this, however, was before Covid-19. The pandemic has hit aviation hard, with a 60% drop in passenger traffic in 2020, and billion-dollar losses for companies.[10] Although air travel is expected to recover as borders reopen and restrictions ease, there is great uncertainty over the future of the industry in a post-Covid world that has learned to work from home, video conference and do without foreign travel, and where long-term health measures (e.g. quarantining) may prevail. The bullish growth projections for the industry made pre-Covid of around 4% per year to 2035[11] are unlikely to come to pass. This could be good news for the climate, but much will depend on how governments respond to the crisis. Already, the 2020 baseline under CORSIA has been shifted to 2019, and additional flexibilities introduced. There is the danger that Covid-19 may paradoxically at best relax the pressure on airlines to curb their emissions and, at worst, trigger active support for an industry that, inevitably, must contract in a net zero world.

Maritime transport has tended to attract less attention from activists and policy makers than aviation, perhaps because the sector mostly concerns freight and is therefore less visible to the general public, and because its complex structure renders the apportioning of responsibility for emissions a genuinely difficult task (Traut et al., 2018). For these reasons, and perhaps also because of decision-making processes that give greater weight to larger shipping nations with little incentive to curb emissions, progress on maritime transport within the IMO has lagged behind even the sluggish pace of ICAO. In October 2018, however, under pressure from the Marshall Islands, the world's third largest shipping registry and also a low-lying island state highly vulnerable to climate change, the IMO's Marine Environment Protection Committee agreed to an "initial strategy" to reduce the average carbon intensity of international shipping by 40% from 2008 levels by 2030, and to cut total annual greenhouse gas emissions from ships by 50% by 2050. Similar to the case of aviation, these developments in the IMO have been both welcomed by some as a step forward, and decried by others as woefully incompatible with the Paris goals (e.g. Gordon-Harper, 2020; Bannon, 2020). At the time of writing, there were concerns that implementation details under development were serving to weaken the initial strategy, prompting the EU to consider bringing shipping under the scope of its ETS, much as it has threatened to do with aviation.

Conclusion: beyond Paris, beyond Covid-19

The world looks very different in the early 2020s to what it did in 2015. In addition to the many developments analysed above, there has been the global shock unleased by Covid-19 and by draconian government intervention, unprecedented in modern peace time, aimed at halting its spread. With economic activity severely curtailed throughout much of the world, and international travel almost at a halt, carbon emissions plunged dramatically in early to mid-2020. Oil prices dropped, temporarily into negative figures, as plummeting demand for transport fuel dramatically exacerbated longer term trends. Emissions rebounded later in the year, but overall, 2020 is likely to have registered a dip in CO_2 emissions of about 7% (Friedlingstein et al., 2020; UNEP, 2020). This drop in emissions, while it would be good news in ordinary times, is not a cause for celebration in this context. The associated huge disruption and hardship is a far cry from the vision of a carefully managed transition to a low-carbon, climate resilient and more equitable world. Unless policy steps are taken, it will also be temporary, with no lasting impact on atmospheric greenhouse gas concentrations or the world's ability to adapt to inevitable climatic changes. More widely, Covid-19 has exposed deep fault lines of inequality and vulnerability, in a world lacking in solidarity, and with no systems in place to manage global shocks effectively.

The economic and social cataclysm triggered by the pandemic and governmental response measures, however, could provide the "creative destruction" needed to break high-carbon path dependency and boost climate resilience. There have been calls from many intergovernmental organisations, such as the International Energy Agency (IEA), the International Monetary Fund (IMF), and the UN, to "build back" or "recover" better in this vein. Many of the world's wealthier governments have set up multi-billion-dollar recovery funds, creating an historic opportunity to nurture the green economy and accelerate decarbonisation, while supporting a just and managed decline of high-carbon sectors that also protects the poorest and most vulnerable. Low fossil fuel prices provide a window of opportunity to remove subsidies and introduce carbon pricing. This opportunity, however, is not being exploited as fully as it could. In general, Covid-19 seems to be reinforcing existing patterns: the more ambitious countries are steering funds towards explicitly low-carbon alternatives (notably the EU, which has earmarked 550 billion euros for green recovery), while the less ambitious are tending to prop up existing fossil fuel-related activities. Across the G-20, almost half of public recovery funds for the energy sector are directed towards fossil fuels, rather than lower-carbon options (SEI et al., 2020[12]).

There are other avenues through which the disruption of the Covid-19 pandemic could trigger positive transformation, notably by "locking-in" lower-carbon behavioural change. The forced expansion of video conferencing may permanently shrink business air travel, while more frequent working from home may cut car-based commuting. Another effect may be to raise public awareness of the potential severity of global threats, while demonstrating that even the most liberal, democratic governments can intervene decisively, speedily, and radically, when faced with what they perceive to be an "emergency". The fiscal, regulatory, and security measures deployed to combat the Covid-19 pandemic are in marked contrast to those (not) put in place following declarations of "climate emergencies" by over 1,800 local and national governments.[13] In the case of climate change, a declaration of a "climate emergency" seems to have largely substituted for action, rather than provoked it.

And this is perhaps the main lesson of the post-Paris period. It is relatively easy to declare policy intent or targets for the distant future – net zero, a climate emergency, even an NDC – but much, much more difficult to take the hard decisions needed to implement it. With the Biden Administration now in charge in the US, 63% of the world's emissions are potentially covered by net zero targets.[14] And yet the history of the climate change regime over the past 30 years should guard against confidence that these will translate to real policy action – to the phase out of coal at home and in overseas investment; to the withdrawal of all financial support for new fossil fuel extraction and exploration; to the cessation of support for air transport and road building; to huge investment in low-carbon urban infrastructure; to a massive funding package for clean energy access in developing countries; to at last giving adaptation the attention it deserves. After all, emissions have risen by over 50% since the world decided to "prevent dangerous anthropogenic climate change" (UNFCCC, Article 2). But a more optimistic interpretation is possible too. There has been a transformation in the economics of decarbonisation, with market trends now firmly in favour of renewable energy, at the expense of coal and oil. Understanding of the climate change challenge has broadly shifted from a more punitive burden-sharing approach, to a common endeavour of pursuing more climate-resilient, sustainable development pathways. Technological development – into hydrogen, green steel, battery storage – is proceeding apace. Evidence of the hazards of a warming world, and not just for poorer countries, is hardening, as witnessed, for example, by the extreme wildfires in Australia (2019), in California (2020), and in the Arctic (both years), along with the devastating floods in Northern Europe (2021) and record-breaking heat over the American and Canadian Pacific Northwest (2021). A growing number of countries, mostly in Europe, are achieving sustained emission reductions. All the pieces are in place for the world's largest emitters to finally match ambitious goals with commensurate policy action. These pieces were arguably already lined up before Paris, but the case for action has hardened even further since then. The opportunities opened up by the tragedy of the Covid-19 pandemic, and by the election of the most supportive US President in history, may finally help to "change the game" and deliver on the promise of the Paris Agreement. The chapters in this book provide an invaluable guide to the challenges that still lie ahead.

Notes

1 E.g. see https://sciencebasedtargets.org/business-ambition-for-1-5c/.
2 Remarks at "What's next after Biden's Summit on the road to COP26", webinar organised by the European Climate Foundation on 26 April 2021. On file with author.
3 https://ukcop26.org/cop26-goals/.

4 https://cambioclimatico.go.cr/press-release-leading-countries-set-benchmark-for-carbon-markets-with-san-jose-principles/.
5 https://climateactiontracker.org/countries/china/.
6 The main exceptions being Iran and Turkey (both among the top 20 emitters), along with governments facing conflict or extreme political instability.
7 US contributions to the core funding of the IPCC and UNFCCC secretariat were either maintained through discreet other channels, or compensated for by other donors.
8 https://corporate.exxonmobil.com/-/media/Global/Files/climate-change/PWT-response-to-Banks.pdf.
9 https://theloadstar.com/marshall-islands-berates-shameful-imo-as-it-ignores-climate-change-threat/.
10 www.iata.org/en/pressroom/pr/2020-11-24-01/.
11 See ICAO website, e.g. www.icao.int/sustainability/Pages/GATO2030.aspx.
12 For updated figures, see www.energypolicytracker.org/.
13 For updated figures, see https://climateemergencydeclaration.org/.
14 See https://climateactiontracker.org/documents/829/CAT_2020-12-01_Briefing_GlobalUpdate_Paris5Years_Dec2020.pdf.

References

Anjaparidze, G. (2019). The Extraordinary Climate Agreement on International Aviation: An Airline Industry. Harvard Project on Climate Agreements. Retrieved from www.belfercenter.org/sites/default/files/files/publication/191021-anjaparidze-viewpoint.pdf.

Arroyo, V. (2018). The global climate action summit: increasing ambition during turbulent times. *Climate Policy*, 18:9, 1087–1093, DOI: 10.1080/14693062.2018.1516957.

Bannon, E. (2020). UN Shipping Body fails to implement its own greenhouse gas reduction plan. Transport and Environment. 23 October 2020. Retrieved from www.transportenvironment.org/press/un-shipping-body-fails-implement-its-own-greenhouse-gas-reduction-plan.

BBC (2018). Climate change: COP 24 fails to adopt key scientific report. By Matt McGrath, 8 December 2018. Retrieved from www.bbc.co.uk/news/science-environment-46496967.

Biniaz, S. (2016). Comma but differentiated responsibilities: punctuation and 30 other ways negotiators have resolved issues in the international climate change regime, *Michigan Journal of Environmental & Administrative Law*, 6, 37–44.

Broberg, M. and Romera, B. M. (2020). Loss and damage after Paris: more bark than bite? *Climate Policy*, 20:6, 661–668, DOI: 10.1080/14693062.2020.1778885.

Climate Action Tracker (2020). Biden's election could bring a tipping point putting Paris Agreement 1.5 degree limit "within striking distance", 7 December 2020. Retrieved from https://climateactiontracker.org/press/bidens-election-could-bring-a-tipping-point-putting-paris-agreement-15-degree-limit-within-striking-distance/.

Climate Change News (2018). Climate science on 1.5°C erased at UN talks as US and Saudis set in. By Sara Stefanini and Karl Mathiesen, 8 December 2018. Retrieved from www.climatechangenews.com/2018/12/08/climate-science-1-5c-erased-un-talks-us-saudis-step/.

Climate Change News (2019). Madrid climate talks to split nations into vanguard and laggard. By Chloe Farand, 21 November 2019. Retrieved from www.climatechangenews.com/2019/11/29/madrid-climate-talks-will-split-nations-vanguard-laggard/.

Climate Change News (2020). Cop25 president Carolina Schmidt blames big emitters for low-ambition climate talks. By Francisco Parra, 15 January 2020. Retrieved from www.climatechangenews.com/2020/01/15/cop25-president-carolina-schmidt-blames-big-emitters-low-ambition-climate-talks/.

Curran, G. (2020). Divestment, energy incumbency and the global political economy of energy transition: the case of Adani's Carmichael mine in Australia. *Climate Policy*, 20:8, 949–962, DOI: 10.1080/14693062.2020.1756731.

Depledge, J. (2006). The opposite of learning: Ossification in the climate change regime. *Global Environmental Politics*, 6:1, 1–22.

Edwards, G., Adarve, I. C., Bustos, M. C. and Roberts, J. T. (2017). Small group, big impact: how AILAC helped shape the Paris Agreement. *Climate Policy*, 17:1, 71–85, DOI: 10.1080/14693062.2016.1240655.

Friedlingstein, P. et al. (2020). Global Carbon Budget 2020. *Earth System Science Data*, 12, 3269–3340, https://doi.org/10.5194/essd-12-3269-2020.

Galik, C. S., DeCarolis, J. F. and Fell, H. (2017). Evaluating the US Mid-Century Strategy for Deep Decarbonization amidst early century uncertainty. *Climate Policy*, 17:8, 1046–1056, DOI: 10.1080/14693062.2017.1340257.

Gordon-Harper, G. (2020). IMO Advances Measures to Reduce Emissions from International Shipping. 24 November 2020. Retrieved from https://sdg.iisd.org/news/.

Grist (2009). Todd Stern's speech cheers delegates at Bonn climate talks. By David Turnbull, 31 March 2009. Retrieved from https://grist.org/article/2009-03-30-todd-sterns-speech-cheers1/.

Guardian (2001). Bush kills global warming treaty. By Julian Borger, 29 March 2001. Retrieved from www.theguardian.com/environment/2001/mar/29/globalwarming.usnews.

Hale, T. N. et al. (2020). Sub- and non-state climate action: a framework to assess progress, implementation and impact. *Climate Policy*, DOI: 10.1080/14693062.2020.1828796.

Hilton, I. and Kerr, O. (2017). The Paris Agreement: China's 'New Normal' role in international climate negotiations. *Climate Policy*, 17:1, 48–58, DOI: 10.1080/14693062.2016.1228521.

Höhne, N. et al. (2017). The Paris Agreement: resolving the inconsistency between global goals and national contributions. *Climate Policy*, 17:1, 16–32, DOI: 10.1080/14693062.2016.1218320.

IPCC (2018). Intergovernmental Panel on Climate Change Special Report on Global Warming of 1.5°C. Masson-Delmotte, V., Zhai, P., Pörtner, H.-O., Roberts, D., Skea, J., Shukla, P. R., … Waterfield, T. https://www.ipcc.ch/sr15/.

IPCC 2021 (2021). Climate Change 2021: The Physical Science Basis. Contribution of Working Group I to the Sixth Assessment Report of the Intergovernmental Panel on Climate Change. Masson-Delmotte, V., P. Zhai, A. Pirani, S. L. Connors, C. Péan,… B. Zhou. Cambridge University Press. In Press.

Kemp, L. (2017). US-proofing the Paris Climate Agreement. *Climate Policy*, 17:1, 86–101, DOI: 10.1080/14693062.2016.1176007.

Kinley, R. (2017). Climate change after Paris: from turning point to transformation. *Climate Policy*, 17:1, 9–15, DOI: 10.1080/14693062.2016.1191009.

Kinley, R., Cutajar, M. Z., de Boer, Y. and Figueres, C. (2020). Beyond good intentions, to urgent action: Former UNFCCC leaders take stock of thirty years of international climate change negotiations. *Climate Policy*. DOI: 10.1080/14693062.2020.1860567.

Kuramochi, T. et al. (2020) Beyond national climate action: the impact of region, city, and business commitments on global greenhouse gas emissions. *Climate Policy*, 20:3, 275–291, DOI: 10.1080/14693062.2020.1740150.

Larsson, J., Elofsson, A., Sterner, T. and Åkerman, J. (2019) International and national climate policies for aviation: a review. *Climate Policy*, 19:6, 78–799, DOI: 10.1080/14693062.2018.1562871.

Lees, E. (2017) Responsibility and liability for climate loss and damage after Paris. *Climate Policy*, 17:1, 59–70, DOI: 10.1080/14693062.2016.1197095.

Lui, S. et al. (2021) Correcting course: the emission reduction potential of international cooperative initiatives. *Climate Policy*, 21:2, 232–250, DOI: 10.1080/14693062.2020.1806021.

Mathews, J. A. (2017). Global trade and promotion of cleantech industry: a post-Paris agenda. *Climate Policy*, 17:1, 102–110, DOI: 10.1080/14693062.2016.1215286.

Mendelevitch, R., Hauenstein, C. and Holz, F. (2019). The death spiral of coal in the U.S.: will changes in U.S. Policy turn the tide? *Climate Policy*, 19:10, 1310–1324, DOI: 10.1080/14693062.2019.1641462.

Monjon, S. and Quirion, P. (2011) A border adjustment for the EU ETS: reconciling WTO rules and capacity to tackle carbon leakage. *Climate Policy*, 11:5, 1212–1225, DOI: 10.1080/14693062.2011.601907.

OECD (2021), Climate Finance Provided and Mobilised by Developed Countries: Aggregate Trends Updated with 2019 Data, Climate Finance and the USD 100 Billion Goal, OECD Publishing, Paris, https://doi.org/10.1787/03590fb7-en.

Olivier, J. and Peters, J. (2020). Trends in Global CO2 and Total Greenhouse Gas Emissions: 2020 Report. PBL Netherlands Environmental Assessment Agency, The Hague.

SEI, IISD, ODI, E3G, and UNEP. (2020). The Production Gap Report: 2020 Special Report. http://productiongap.org/2020report.

Setzer, J. and Higham, C. (2021) Global Trends in Climate Change Litigation: 2021 Snapshot. Grantham Research Institute on Climate Change and the Environment and Centre for Climate Change Economics and Policy, London School of Economics and Political Science: London.

Sharma, A. (2017). Precaution and post-caution in the Paris Agreement: adaptation, loss and damage and finance. *Climate Policy*, 17:1, 33–47, DOI: 10.1080/14693062.2016.1213697.

Streck, C. (2020). The mirage of Madrid: elusive ambition on the horizon. *Climate Policy*, 20:2, 143–148, DOI: 10.1080/14693062.2020.1726564.

Traut, M., Larkin, A., Anderson, K. McGlade, C., Sharmina, M. & Smith, T. (2018) CO2 abatement goals for international shipping, *Climate Policy*, 18:8, 1066–1075, DOI: 10.1080/14693062.2018.1461059.

Viñuales, J. E., Depledge, J., Reiner, D. M. and Lees, E. (2017). Climate policy after the Paris 2015 climate conference. *Climate Policy*, 17:1, 1–8, DOI: 10.1080/14693062.2016.1242060.

UNEP (2020). Emissions Gap Report. Nairobi. Retrieved from www.unenvironment.org/emissions-gap-report-2020.

UNEP/Ozone (2019). Kigali Amendment Implementation Begins. Retrieved from https://ozone.unep.org/kigali-amendment-implementation-begins.

UNEP/SCCL (2020). Global Climate Litigation Report: 2020 Status Review. United Nations Environment Programme and Sabin Center for Climate Law. Retrieved from https://wedocs.unep.org/bitstream/handle/20.500.11822/34818/GCLR.pdf?sequence=1&isAllowed=y&mc_cid=2ee16a4b8f&mc_eid=5ba5be7c40.

Climate policy after the Paris 2015 climate conference

JORGE E. VIÑUALES, JOANNA DEPLEDGE, DAVID M. REINER, AND EMMA LEES

The landmark 2015 Paris Agreement, adopted to widespread acclaim at the 21st Conference of the Parties (COP 21) to the United Nations Framework Convention on Climate Change (UNFCCC), reflects profound political cleavages and tensions, which were already condensed, even camouflaged, in the wording of the mandate that launched the negotiations, as set out in the 2011 Durban Platform for Enhanced Action.[1] Of particular note are the opaque formulations defining both the mandate assigned to negotiators to work on 'mitigation, adaptation, finance, technology development and transfer, transparency of action, and support and capacity building' and the target outcome, potentially consisting of a 'protocol, another legal instrument or an agreed outcome with legal force under the Convention applicable to all Parties'.[2] Writing in hindsight, the authors of this Editorial would have hoped to be able to comment on how these many tensions were settled and perhaps solved. But this is, alas, not possible, at least not yet. Rather than settled, the tensions arising in many – perhaps most – major issues, ranging from mitigation to adaptation to the different avenues to promote implementation (above all finance) were instead ingrained into the text of the Paris Agreement, leaving them to be resolved (perhaps) through the instrument's future operation.

The wide diversity of possible interpretations of so many different parts of the Paris Agreement is clearly reflected in the contrasting views expressed in the articles of this guest-edited issue of Climate Policy. The range of interpretations was even wider in the conference held at the University of Cambridge, on 22 January 2016,[3] from which several of these articles are derived. Such diversity may in turn be diversely interpreted. But in the opinion of the authors of this Editorial, this is worrying, because it has the potential to obfuscate, at least for now, attempts at answering the main question raised by COP 21: how confident can we be that the Paris Agreement has set the world on a path that will eventually avoid dangerous climate change?

Not all the diverging views expressed and discussed at the Cambridge conference[4] – and indeed in the wider commentaries and debates that have emerged since Paris – can be addressed in one guest-edited issue. Before turning to those that are indeed encompassed here, it is worth highlighting one key tension that is hidden in both the wording of the Paris Agreement and the difference between the two global temperature targets referenced (albeit in different ways) in that agreement, namely,

2°C and 1.5°C. Both targets had been mentioned in international climate negotiations since at least the 2009 Copenhagen Accord.[5] However, the lion's share of the research effort has focused on the 2°C target (e.g. see Jordan et al., 2013), so that the implications of climate stabilization at 1.5°C – or at least attempting to do so – are poorly understood. As we write this Editorial, many groups of scientists and modellers are actively working on filling this important gap in our knowledge. Holding the increase of global average temperature to 'well below' 2°C as compared with pre-industrial times and 'pursu[ing] efforts' to limit the increase to 1.5°C, as called for in the Paris Agreement, may potentially have dramatic socio-economic consequences. Behind the compromise wording used in Article 2(1)(a) may lie either major consequences for low-lying small island nations (in case of a higher increase) or (in case of drastic limitations) the scrapping of massive investments in infrastructure with the ensuing socio-economic, political, and strategic impacts. In urging further research on the implications of a 1.5°C target,[6] decision 1/CP.21 adopting the Paris Agreement has greatly influenced the scope of research efforts. These should include not just studies of impacts and mitigation costs, but assessment of the feasibility of the required political and societal response, on anything like the timescale needed.

Should there be any doubt, from a policy perspective, that the (aspirational) 1.5°C goal will indeed help set the world onto a safer climatic path? Although ambitious, such an aspirational goal may or may not add momentum. If a target is perceived as unrealistic, particularly after its socio-economic implications have been estimated, will this help or undermine the normative pull of the Agreement? At the Cambridge conference, some speakers suggested that a 1.5°C target might be incompatible with democracy to the extent that it would require such a large-scale and intrusive intervention that it could strain democratic decision-making processes. On the other hand, limiting ambition to a 2°C target may entail severe consequences – actual 'loss and damage' – for a number of countries and populations. Such major tensions lie beneath the laconic expressions 'well below' and 'pursue efforts' in Article 2(1)(a), which are but two among many other illustrations of how tensions have been 'ingrained', rather than addressed, in the text of the Paris Agreement.

A wide divergence of views is also apparent in many topics analysed in this guest-edited issue. The highly optimistic stance adopted by Richard Kinley, Deputy Executive Secretary of the UNFCCC, in the first article can be contrasted, in this respect, with the disquiet expressed in Anju Sharma's analysis of the deal reached on adaptation, finance, and loss and damage, or with the optimism for the Agreement's future expressed – based on a different rationale – in the piece by Niklas Höhne and colleagues on Intended Nationally-Determined Contributions (INDCs).[7] Similarly, the article by Emma Lees views the new provision in the Paris Agreement on loss and damage in a clearly different light, when compared to Sharma's perspective on this question. Reasonable minds may certainly disagree and that is part of academic debate. What is more worrying is that such disagreements – and those expressed during the Cambridge conference – may reflect the potential for reasonable disagreements among countries and other stakeholders as to the path envisioned by the Paris Agreement. The international community faced a similar problem in the wake of the 1992 UN Conference on Environment and Development in Rio de Janeiro (Viñuales, 2013). Here, the broad concept of 'sustainable development' allowed such a wide array of views and positions to be brought under an apparently single and simple umbrella that, as international efforts sought to move from norm development to actual implementation, this led to a new set of no less than 17 Sustainable Development Goals (SDGs), with 169 targets and potentially many more indicators. Such an expansive definition of sustainable development has certainly helped to secure consensus, but more by 'papering over' substantive differences, rather than truly resolving them. The risk, of course, is that governments are on board with the process, but the difficult decisions needed for real change are never taken. As with the Paris Agreement, only time can tell whether the SDGs will set humanity on course towards truly 'sustainable' development and, as with the Paris Agreement, the jury is likely to be out for a long time.

In the first article of this special issue, which draws on the opening keynote speech he gave to the Cambridge conference, Richard Kinley offers the unique insight of someone who has been at the centre of the climate negotiations since the early 1990s. Kinley considers COP-21 as 'the most successful climate conference ever', highlighting several major developments that took place both within and outside the plenary halls, as well as eight features of the outcome that, he claims, have 'changed the game' towards a more cooperative and effective approach to climate governance. Kinley's article can usefully be considered as a platform, a line drawn by someone who has been in the 'game' since the beginning.

Kinley's optimism on the provisions of the Paris Agreement relating to mitigation and innovations in climate governance is echoed in the arguments put forward by Niklas Höhne and his colleagues. But the reasons given by Höhne et al. differ in critical ways. While Kinley is broadly optimistic about the Paris Agreement as it stands, Höhne et al. see optimism in the potential to strengthen both the treaty, and country INDCs, in the (hopefully near) future. Indeed, their analysis of the INDC system eventually retained in the Paris Agreement looks at the still significant gap between the Agreement's long-term global goals and the overall impact of the INDCs as a glass half-full, rather than half-empty. They see cause for hope in several factors, notably, the fact that many INDCs, including those of some large emitters, were conservative, leaving significant room for easy improvement, and the positive institutional dynamics that were set in motion at the national level by the very process of preparing the INDCs. Höhne et al. are similarly optimistic about the mechanisms introduced by the Paris Agreement to increase ambition over time, and the important role that 'non-party stakeholders' are likely to play in future mitigation efforts. They also point to some observed or projected technological transformations that could facilitate low-carbon trajectories, such as the development of renewable energy technologies and electric cars. Höhne et al. claim that the success of mitigation action will depend firstly on the prompt and full implementation of (soon to be) nationally-determined contributions (NDCs), involving non-party stakeholders, and then on harnessing technological transformations to *overachieve* goals and hence generate momentum for the wider economy to switch to a new production matrix. There is some convergence of views, then, that the Paris Agreement has steered the world onto a path that can avoid dangerous climate change, *so long as its promise to trigger stronger future action is realized.*

A key area of the Paris Agreement where there is a marked divergence between the optimism of Kinley and the much more sober interpretations expressed by Anju Sharma, concerns adaptation, finance, and loss and damage. Sharma asks whether, aside from what was politically feasible, the results of COP-21 in these areas are satisfactory from the perspective of the most vulnerable constituencies. Through a detailed analysis of the provisions of the Paris Agreement, Sharma argues that these do not constitute an adequate response to the likely impacts of climate change, whether from a precautionary (anticipative) or post-cautionary (redress) perspective. Sharma pays particular attention to questions of finance and notes a number of shortcomings in both the Paris Agreement and the accompanying decision 1/CP.21, including the fact that there are no new financial commitments under the Paris Agreement, with the timeframe for the US$ 100 billion pledge agreed in Copenhagen simply extended to 2025. At the same time, she points to the lack of reference to 'new and additional' funds in the Paris Agreement and continued controversies over the sources (public or private) of funds. The imbalance between the funds directed to mitigation as compared to those for adaptation is also a cause of concern, as is potential over-counting in climate finance reports of the funds that have genuinely been directed towards adaptation. Sharma further cautions that, despite the enthusiastic reactions of many observers regarding the enhanced status of adaptation in the Paris Agreement, the relevant provisions 'fall well short' and the provisions on climate finance 'do not inspire confidence'. Sharma argues that, from the perspective of the world's poorest communities, the Paris Agreement has not brought us much closer to avoiding dangerous climate change in a fair and equitable manner, if at all.

On the topic of loss and damage, Emma Lees expresses some cautious optimism regarding the inclusion of a specific provision in the Paris Agreement itself (Article 8). Concerns about the potential use of this provision to claim liability and compensation for loss and damage led to the addition of paragraph 51 in the accompanying COP decision, according to which Article 8 'does not involve or provide a basis for any liability or compensation'. Relying on general concepts of private law, Lees analyses these provisions in the wider context of the UNFCCC and the Paris Agreement and argues that 'responsibility' must be clearly distinguished from 'liability and compensation', the latter being one potential consequence (among others) of the former. According to Lees, the loss and damage provision could be fleshed out not only through insurance mechanisms, which seems to be the politically easier method, but also through a differentiated allocation of responsibility that does not necessarily trigger liability and compensation (and is hence consistent with paragraph 51). In fleshing out legally this allocation, two key questions arise, namely how to fine-tune traditional concepts of causation and fault to make them adequate for the specificities of climate change and what consequences (other than liability and compensation) could be derived from such an allocation. Lees cautions against ignoring the current ambiguities surrounding the loss and damage provisions, which may result in depriving this potentially important dimension of the Paris Agreement of any concrete effect. Even if the Paris Agreement has increased our chances of eventually stabilizing the climate at a 'safe' level, some 'loss and damage' due to climate change over the coming decades is inevitable.

Setting aside, for now, concerns over its substance, there is no doubt that merely reaching a consensus on the Paris Agreement was a major diplomatic achievement. The politics of Paris clearly played out very differently from those at COP 15 in Copenhagen six years before, which ended in widespread acrimony without a consensus outcome (e.g. see Depledge, 2010). How did this come about? In his article, Kinley praises the 'strong leadership (not least within the G-77 and China) and flexibility' that was demonstrated 'by all of the key players'. Isabel Hilton and Oliver Kerr's article focuses on a central player in the climate negotiations, namely China. They analyse in depth what they see as the more cooperative attitude of the world's largest emitter, which they trace back to a change in China's economic development model starting in the 12th Five Year Plan Period (2011–2015). They highlight China's efforts to 'move away from energy-intensive industry, exports and investment-led growth, and toward services and domestic consumption, with a focus on low carbon technologies', which, according to them, helps to explain the redefinition of its agenda in international climate negotiations. This, in turn, accounts at least in part for the more constructive role that China played in the negotiations that led to the Paris Agreement, as compared with its more reticent stance throughout the Copenhagen process. Of particular significance, here, was China's emerging partnership with the US, cemented through a series of coordinated policy announcements leading up to COP-21. These political initiatives in effect brought to an end the unproductive 'you go first' relationship between the world's two highest emitters that had frustrated the negotiations for decades. China's rising emissions and growing confidence on the world stage, and a US President willing to use political and diplomatic capital in the climate change arena, enabled both countries to move together, thereby dropping a critical piece of the political puzzle into place. China's prompt ratification of the Paris Agreement, again in a joint move with US President Obama at the September 2016 G20 Summit in Hangzhou, confirms China's more proactive engagement with the climate negotiations and ensures this will be sustained into the future. While China's apparently strong support for the Paris Agreement is not a sufficient condition to avert dangerous climate change, it is certainly a necessary one, and therefore a cause for cautious optimism.

The increasingly constructive collaboration between China and the US on climate change is perhaps the single biggest political factor accounting for the success of COP-21. While Hilton and Kerr document China's critical role in this, much credit must also be given to President Obama who expended considerable time, energy, and resources in building up a strong working relationship with China. To recall, this was a huge turnaround from the frosty encounters in Copenhagen. President Obama has now taken his pro-climate position to its logical conclusion, and formally 'accepted'[8] the Paris Agreement as an Executive Agreement, as announced at the 2016 G20 Hangzhou Summit. This did not require advice and consent of the US Senate, which would have been impossible to obtain, given the constitutional hurdles involved, coupled with continuing US exceptionalism on the whole issue of climate change. Indeed, a critical question throughout the Paris negotiations was precisely how to craft the outcome so as to enable the US to become a party while bypassing the Senate: this requires linkage to existing US legislation (such as the US Clean Air Act), an existing treaty which had been ratified by the Senate (notably, the UNFCCC) and/or the President's own powers under the US Constitution. While there is rejoicing now at US acceptance of the Paris Agreement, the drawback of Executive Agreements are that they are vulnerable to being revoked by future Presidents. President Obama was supportive of climate change action, but what will be the views of his successor, or indeed future US Presidents down the line? The Paris Agreement is a long-term instrument, whose effectiveness (as argued by Höhne et al.) will depend on its future strengthening. It is entirely possible, however, that a future US President would revoke its membership of the treaty. These important questions are discussed in detail by Luke Kemp, in his article on 'US-proofing' the Paris Agreement. Kemp argues that the Paris Agreement is highly vulnerable to withdrawal by the US (or indeed any other major party, as Canada did from the Kyoto Protocol), not least because it does not include any provisions relating to non-parties. He suggests possible amendments to the Paris Agreement that could guard against dropouts, while also exploring possible responses in the event of withdrawal by a major party. In doing so, his aim is to ensure continued effective participation by entities from the withdrawing party, such as sub-national governments and businesses.

And of course, China and the US were far from being the only participants whose constructive diplomacy in Paris helped to seal the deal. According to Kinley, there was 'a strong, almost palpable, determination by all governments to reach agreement'. Attracting the most media attention was perhaps the High Ambition Coalition, eventually bringing together upwards of 100 Parties, from the richest to the poorest, who raised the bar of acceptable action, and forced through 'high ambition' provisions such as the 1.5°C target and the specific provision on loss and damage.

Lower profile groups, such as the Independent Association of Latin America and the Caribbean, or AILAC,[9] also had important impacts on the outcome. This coalition of small and middle-sized countries, formed in 2012, displayed a remarkable capacity for bridge-building and leading by example. Guy

Edwards, Isabel Cavelier Adarve (former head of the AILAC support unit, writing in a personal capacity), Maria Camila Bustos, and J. Timmons Roberts, document the rise and influence of this 'small group with the big impact', arguing that AILAC was critical to consensus building in Paris. They explore how AILAC helped to shape the Paris Agreement, and ask how it can build on its success to continue to nudge the climate regime along the path of a climatically safe future. Although the spotlight tends to fall on larger emitters in climate change politics, nearly a third of global emissions are accounted for by small and middle-sized countries. Given that the short-term prospects for decisive policy action in several of the largest countries and regions – US, India, Russia, even the EU – are lukewarm at best for diverse political reasons (even the promising Chinese resolve remains to be tested), the actions of small and middle-sized countries such as AILAC may become critical in maintaining the Paris Agreement's momentum. There are no sanctions in the Paris Agreement to enforce implementation and no automatic mechanism to strengthen its targets (only processes to facilitate this). Peer pressure, leadership by example and insistent scrutiny by progressive governments (not just NGOs) are therefore the central (perhaps only) means by which the promise of the Paris Agreement can be fulfilled. Edwards and his colleagues have shown how a group of small and middle-sized countries can exert positive influence on the negotiations; the hope is that they will now exert the same kind of positive force on implementation and strengthening.

These constructive political developments are surely one of the main reasons to be optimistic about the Paris Agreement's prospects: that its negotiation and adoption brought countries together in a spirit of constructive cooperation and support for the multilateral process that had eluded the climate change regime for many years. But a positive political atmosphere does not, by itself, cut emissions and avoid dangerous climate change. For that, implementation is needed. The speed with which countries ratify a treaty, and therefore with which it enters into force, can provide an early signal of commitment to put into place the necessary laws, policies, and measures to implement its provisions. In this respect, the Paris Agreement would seem well placed. The leadership shown by China, the US, many small island states and others in declaring their early commitment to the Paris Agreement, led to a flood of ratifications triggering entry into force of the Agreement on 4 November 2016, less than one year after its adoption. This is remarkable, beating even the rapid entry into force of the UNFCCC itself (22 months). One could be cynical, of course, and argue that it is easy for governments to ratify the Paris Agreement, when all they are doing is confirming their own pre-existing policies as set out in their INDCs, which are, in any case, voluntary.

Indeed, the voluntary and inclusive approach of the Paris Agreement, extending non-legally commitments to all countries, represents a significant departure from the now out-of-favour approach of the Kyoto Protocol, with its legally binding targets for a narrower group of countries. Assessment of the effectiveness of the Kyoto Protocol remains divided – high levels of compliance, indeed over compliance, were achieved. However, use of the flexibility mechanisms (such as the CDM) and accounting rules for land-use change and forestry played an important part in this (for an in-depth assessment, see Shishlov, Morel, & Bellassen, 2016). There is also the view that the Kyoto Protocol was 'designed for obsolescence' because of its firewall between Annex I Parties with emission commitments, and non-Annex I Parties without (Reiner & Jacoby, 2001). There can be no doubt, however, that the legally binding framework of the Kyoto Protocol helped to promote compliance, and exerted pressure on governments to respond with a degree of regulatory attention that may not be reproduced for voluntary commitments. The new approach enshrined in the Paris Agreement will therefore need to prove its worth, and that proof will lie in actual implementation through concrete emission reductions.

One issue that will undoubtedly come to the fore as countries seek to strengthen their climate policies to implement the Paris Agreement is the relationship of those policies, and indeed the Agreement itself, with trade rules under the World Trade Organization (WTO). The potential for clashes between climate-related goals and free trade rules is one that has concerned academic commentators for decades (e.g. Esty, 1994). In his article on global trade and the promotion of cleantech industry post-Paris, John A Mathews argues that 'these disputes have come to a head in the past few years… imposing a serious risk of reversing whatever gains might be made in the forums established within the terms of the UNFCCC'. Mathews calls for a reconciliation between the WTO and the climate regime, which could be done through a COP resolution calling on the WTO to recognize climate change mitigation as a 'public good', whose pursuit might justify the application of GATT Article 20 exemptions in specified circumstances and for a designated period. This would have the advantage, for example, of giving greater flexibility to countries to develop their own clean industry capabilities, with less fear of triggering trade disputes. The extent to which the global trading regime can be harnessed in explicit favour of climate protection, rather than as a perceived threat or sometimes actual (unwitting) obstacle to it, will greatly impact on the effectiveness of the Paris Agreement in, as Höhne et al. put it, 'flipping' global markets towards the low-carbon technological future that is needed to achieve the 2° goal.

Overall, the articles in this guest-edited issue provide both a retrospective and a prospective view of the main outcomes of COP-21 and their policy implications. Most of all, they signal the wide range of interpretations of an instrument that, rather than reflecting consensus on the path forward, seems to encapsulate different – widely different – potential paths. To return, then, to the question posed at the beginning of this editorial: the extent to which the Paris Agreement has set us on a course to avoid dangerous climate change will depend on the path that is eventually taken. It is clear that the Paris Agreement, and the INDCs within it, is not sufficient as it stands to ensure a safe and stable climate. It does, however, prop open the window for doing so, and provide a politically acceptable structure and process through which *all* countries can play their part and tighten their commitments over time. If the Paris Agreement had not been concluded, and COP-21 had ended in the same kind of diplomatic debacle as Copenhagen, then this would surely have sounded the death knell for the international climate change regime. The euphoria felt after Kyoto in 1997, and the gloom following Copenhagen, have now been replaced with a cautious optimism after COP-21. The Paris Agreement has taken climate governance in a new direction, which holds both promise and pitfalls; as noted by Richard Kinley, the game is indeed changing. We hope that the articles in this collection will help illuminate this brave new world of the Paris Agreement, and the potential pathways to follow for winning that game.

Disclosure statement

No potential conflict of interest was reported by the authors.

Notes

1. 'Establishment of an Ad Hoc Working Group on the Durban Platform for Enhanced Action', Decision 1/CP.17, 15 March 2012, doc. FCCC/CP/2011/9/Add.1, 2.
2. Decision 1/CP.17, para 2.

3. The conference was jointly organized by four different centres within the University of Cambridge: the Centre for Environment, Energy and Natural Resource Governance (C-EENRG), the Energy Policy Research Group (EPRG), the Department of Politics and International Studies, and the Centre for Rising Powers, in association with Climate Policy.
4. For an anonymized summary of the conference proceedings, see R. Rafaty, 'What now? Climate change and energy after Paris', Policy Briefing, available at http://www.ceenrg.landecon.cam.ac.uk/ (visited on 24 July 2016). The website includes additional materials from the conference including many of the presentations.
5. 'Copenhagen Accord', Decision 2/CP.15, 30 March 2010, Doc. FCCC/CP/2009/11/Add.1, paras 1, 2, and 12.
6. "Adoption of the Paris Agreement", Decision 1/CP.21, 29 January 2016, Doc. FCCC/CP/2015/10/Add.1, para 21 (this paragraph is addressed to the IPCC, but the IPCC will necessarily rely on available research work done by the scientific community, and such work is for the time being scarce).
7. Takeshi Kuramochi, Carsten Warnecke, Frauke Röser, Hanna Fekete, Markus Hagemann, Thomas Day, Ritika Tewari, Marie-Jeanne Kurdziel, Sebastian Sterl, and Sofia Gonzales.
8. An act that under both US constitutional law and international law triggers the same effects as ratification.
9. Asociación Independiente de América Latina y el Caribe in the original Spanish.

References

Depledge, J. (2010). The outcome from Copenhagen: At the limits of global diplomacy. *Environmental Policy and Law, 40*(1), 17–22.

Esty, D. C. (1994). *Greening the GATT: Trade, environment, and the future*. Washington, DC: Peterson Institute.

Jordan, A., Rayner, T., Schroeder, H., Adger, N., Anderson, K., Bows, A., ... Whitmarsh, L. (2013). Going beyond two degrees? The risks and opportunities of alternative options. *Climate Policy, 13*(6), 751–769.

Reiner, D. M., & Jacoby, H. D. (2001). Learning the lessons of Kyoto. *Climate Policy, 1*(2), 273–275.

Shishlov, I., Morel, R., & Bellassen, V. (2016). Compliance of the parties to the Kyoto protocol in the first commitment period. *Climate Policy, 1*(2), 273–275.

Viñuales, J. E. (2013). The rise and fall of sustainable development. *RECIEL, 22*(1), 3–13.

Climate change after Paris: from turning point to transformation

RICHARD KINLEY

COP 21 was the most successful climate change conference ever. This article lays out eight key ways in which the Paris Conference changed the game: the shift in emphasis towards national action; cementing of the below 2 °C goal and going further to reference 1.5 °C; the addition of 'global peaking' and balance between emissions and removals as supplementary goals; global stock-taking every five years with no backsliding; much more nuanced differentiation between developed and developing countries; greater recognition of adaptation and the inclusion of loss and damage; new developments on finance, including stronger reporting provisions; and renewed recognition of market mechanisms.

Policy announcements and initiatives made outside of the formal negotiations were also spectacular in scale and scope, suggesting that a new sustainable growth model is underway. At the same time, non-state actors are becoming the engine of both mitigation and adaptation action. There are several reasons for the success of COP 21, including the positive underlying economic realities, a strong determination to succeed on the part of all governments, outstanding diplomacy on the part of the French Presidency and the realization that all must participate in solving the huge global challenge posed by climate change. The game is not over, however. Targets now need to be exceeded and strengthened, without waiting for the formal review processes. Implementation rules must be developed and the pledged support for developing countries must be actioned. The Paris Agreement provides the directional signal, but it is through its implementation that its ultimate success will be judged.

Policy relevance

This article, which presents the perspective of a senior international climate change official, will assist policy makers in better understanding the Paris Agreement and its achievements, along with wider dynamics in the international climate change regime.

The twenty-first Conference of the Parties[1] to the United Nations Framework Convention on Climate Change (UNFCCC) in Paris – COP 21 – is recognized as the most successful climate change conference ever. A record-breaking 28,000 participants entered the premises at Le Bourget[2] over the two weeks of the event; they were well taken care of and enjoyed very smooth logistics. Even more significantly, the first day of the conference saw the largest ever number of heads of state and governments together under one roof on one day and all who wished to speak could do so, providing a huge political push for success[3]. A vast array of side events and announcements provided ample demonstration of the momentum and engagement towards climate solutions.

There was a remarkable spirit of cooperation and a determination among governments to reach agreement, enabled by outstanding diplomacy by the Presidency under the leadership of Laurent Fabius, France's then Minister of Foreign Affairs and International Development. The crowning achievement was the adoption of the Paris Agreement and its related decisions (UNFCCC, 2015a), which constitute a turning point in global efforts to deal with the climate change problem and chart a new course to a low-carbon, climate-resilient future.

The phenomenon that was COP 21 constitutes a ground-breaking outcome. Notwithstanding some dissenting voices, the adjectives that have flowed since the 12 December 2015 reflect the widespread euphoria of the moment and are markedly different from all COPs since Kyoto in 1997: 'historic', 'ambitious', 'monumental triumph', 'landmark' (BBC, 2015; New York Times, 2015; UN News Centre, 2015). And for the first time in history, the results of a COP actually exceeded the expectations of virtually all forecasts. Among the many achievements of the Paris meeting, there are perhaps eight that stand out:

1. Climate change mitigation is now firmly founded on national action; the orientation has shifted from a perceived top-down one, à la Kyoto Protocol, to one with a strong bottom-up component based on national undertakings. The submission before Paris of Intended Nationally Determined Contributions (INDCs) – now set to become NDCs – in numbers that no one ever contemplated[4] affirms this new reality. Importantly, these INDCs go beyond mitigation, also covering adaptation, finance and other issues, thus reflecting more fully the broader realities involved in acting on climate change
2. The global goal of limiting temperature change to below 2 °C compared with pre-industrial times had been agreed five years ago. In Paris, governments enshrined this in a legal treaty and went further by agreeing to 'pursue efforts' (UNFCCC, 2015a, Article 12.1a) to limit the increase to 1.5 °C. This constitutes an important statement of solidarity with vulnerable countries; it may be an aspirational goal, but aspirations create expectations and are thus extremely important in politics.
3. The inclusion of a more popularly understandable long-term goal, alongside the below 2 °C global temperature goal, was the most significant achievement. The inclusion in a legal Agreement of the concepts of global peaking of emissions and a balance between global emissions and removals (in other words, zero net emissions) means that the policy certainty that the business sector has been seeking has been provided. Although the timelines are somewhat open, the signal is clear that the era of fossil fuels is ending. Real transformation of the energy sector is the will, and undertaking, of all of the world's governments, including fossil-fuel exporters.
4. Linked to the long-term goal, the agreement on a process of global stock-taking every five years 'in the light of equity and best available science' (UNFCCC, 2015a, Article 14.1) will systemize moving to more ambitious action. Again, the direction is clear, with an expectation of continuous improvement (no backsliding). There is a new stability in the international process. It is also significant that this stock-taking will apply not only to mitigation, but also adaptation, as well as means to of implementation and support for developing countries.
5. The international architecture has shifted from a situation of strong differentiation between industrialized and developing countries to one that is much more nuanced. These nuances were possible

because governments realized that enhanced action by all was needed. Differentiation is still woven systematically throughout the Agreement (and in extremely astute ways), while the era of a single, universal system – common commitments with flexibility, reflecting national realities – is also launched. Central to this success is the significantly enhanced attention to the need for support to developing countries to enable them to do more. There is a common direction of travel, but industrialized countries must lead the way.

6. Adaptation has come into its own. One could debate whether parity with mitigation has been achieved, but adaptation is now a pillar of the international climate change regime. Loss and damage was also granted its own article in the Agreement, responding to a key demand of the most vulnerable countries.
7. The issue of finance – or means of implementation – was, as expected, hotly contested and long-standing tensions remain. However, a deal was struck, to the relief of many. The US$ 100 billion financing goal from the 2009 Copenhagen Conference (COP 15) remains in place and will be enhanced from 2025, but without a defined pathway to its achievement. More noteworthy are three new developments:
   - Significantly enhanced attention to the reporting and assessment of support to developing countries both as part of the transparency system and as fundamental conditions of the success of the Agreement
   - Recognition (for the first time) of the possibility of developing countries providing such support (while reaffirming developed countries' existing financial obligations)
   - A surprisingly robust outcome on capacity building to enable developing countries to fulfill their undertakings
8. Market approaches found their way into the Paris Agreement, giving them a renewed role in international efforts to combat climate change. For many, this was another surprise that emerged from COP 21. Article 6 speaks of 'internationally transferred mitigation outcomes' to achieve NDCs and establishes a new 'mechanism' to contribute to mitigation and support sustainable development; the details must now be worked out. . These market approaches are accompanied by a 'framework for non-market approaches to sustainable development', (UNFCCC, 2015a, Article 6.9) whose details will now also be developed.

And on the sidelines of COP 21 hugely important development initiatives were announced, including those on renewables for Africa (UNFCCC news, 2015b), solar energy (UNFCCC news, 2015a) and billions in additional contributions (UNFCCC news, 2016) earmarked for important small funds[5], to mention a few examples.

Alongside the Paris Agreement, it is important to highlight another paradigm shift. Over the past two years the world has experienced an exponential growth in climate action by cities, regions, businesses and civil society, in addition to national actions[6]. On the one hand, this reality was inspired by the ambitions for Paris and mobilization for the global extravaganza that was COP 21; on the other hand it was also an enabler of the ambitious outcomes of the Paris meeting.

The scope and scale of the initiatives and announcements during COP 21, some at the highest level, are spectacular[7]. With the adoption of the Paris Agreement, and the directional signal provided by the long-term goal, we can now look forward to this trend accelerating through a new sustainable growth model that is becoming a virtuous circle of action. Although it is true that the profile of non-state actors in the Paris Agreement was lower than many had hoped for[8], they are increasingly becoming the engine of both mitigation and adaptation action. This is helping to define a 'new normal'. For example, actions by investors are driving a new economy (e.g. a record US$ 330 billion in new clean-energy investment in 2015; Bloomberg, 2016). Similarly, disinvestment initiatives (with managers of US$3.4 trillion in assets acting) and the closing of coal-fired power plants (DivestInvest, 2016) are changing the energy paradigm. The rapidly declining price of renewable energy[9] and the explosion of mega-sized solar and wind power installations[10] are game changing realities. These are but a few examples of a much wider range of mobilization and engagement initiatives.

One cannot consider the results of COP 21 without addressing questions relating to the process and procedure. Who can remember a smoother COP, with so few procedural disputes? To what can one ascribe this success, verging (in this author's opinion) on miraculous? There are several reasons, founded on the desire of governments to reach an agreement. The importance of the presence of heads of state and governments on the first day, and their resounding calls for agreement and ambition, cannot be overstated. It was very helpful to have a clear and agreed timetable from the outset – especially when the preparatory Ad Hoc Group on the Durban Platform (ADP)[11] would close – followed by an efficient transition to French leadership. Mindful of problems experienced by the Danish Presidency of the Copenhagen COP in 2009, the COP President was particularly clear on his intentions regarding negotiating texts. First that there was no 'French text' (and there really was none), but only text derived from Party consultations. The final text emerged in three pre-announced iterations, carefully and often elegantly constructed with – as Fabius promised – 'no surprises'. One can only hope that this experience will put to rest the idea that governments can actually negotiate intricate climate change agreements line by line amongst themselves; negotiation through the Chair's iterations is surely the only viable option for such highly complex discussions as those in Paris. The French presidency was very transparent, consulting all groups (not just the powerful), not imposing its own ideas, working closely with Peru (the hosts of COP 20 the previous year) as a partner, enlisting the support of key negotiators in their team and truly listening to what Parties were saying.

There are many reasons why the COP 21 achieved what it did; four stand out as the main contributors.

First, underlying economic realities such as the spread of renewable energy noted above have shifted to the point where real change towards a lower-carbon world is now possible.

Second, there was a strong, almost palpable, determination by all governments – evident even on the first day – to have an agreement. Ample opportunities to block were not exercised; 'red lines' changed to 'green lines' and a spirit of compromise really did prevail. Strong leadership (not least within the G-77 and China) and flexibility were demonstrated by all of the key players on issues such as adaptation, loss and damage, differentiation and finance. And the leaders of the most powerful, as well as the most vulnerable, countries were personally involved in COP 21 or active behind the scenes.

Third, it is worth reiterating the role played by France (the host country), driven by the COP President, Laurent Fabius – the hospitality, the astute political leadership and effective diplomacy over many months, the listening, the honest-broker role, the willingness to push the envelope and insert ambitious, nuanced language into the Agreement. All of these factors enabled everyone to come away feeling that they had won something.

Fourth, perhaps the most fundamental enabler of the success of Paris was the realization that all must participate in solving the huge global challenge posed by climate change. The seriousness and urgency of the problem demanded nothing less, and the Paris Agreement was only possible because of the contribution of numerous actors in the run-up to COP 21 and at the conference itself.

So, the game has changed, but is it in the bag and problem solved? Here one has to shift from euphoria to sombre assessment. In real terms, the game is actually only just beginning. To some extent, Paris has been deemed such a success precisely because progress in the negotiations since Kyoto has been so slow and so hard won. As Bill McKibben (of 350.org fame) has said: 'the most compelling thing you can say about Paris is not that it saved the planet, but that it saved the chance of saving the planet' (Hood, 2016). We must beware of complacency, and actually need to be concerned that momentum is maintained and not lost. The task ahead is huge, in many ways.

The INDCs, although quasi-universal, are inadequate to the task. They are projected to get the world to somewhere near 3 °C of warming (UNEP, 2015; UNFCCC, 2015b). Although this is a major shift of the trajectory of emissions, these INDCs need to be exceeded and strengthened, without waiting for the stock-taking in 2023, or even the 'facilitative dialogue' planned for 2018.

Global emissions are still rising. The impacts of climate change are going to increase in frequency and magnitude. The Agreement calls for global peaking as soon as possible; and this must come in the next 5–10 years.

The intergovernmental process has a huge task ahead in fulfilling the mandates in the Agreement and the COP decision – developing modalities and rules, getting institutions off the ground or shifted into a higher gear and putting the transparency regime in place. The effort begins with encouraging prompt ratification by governments, to ensure that the Agreement enters into force quickly.

The promise of enhanced support for developing countries has to be fulfilled or much of the Agreement's promise will be lost. Such support will be crucial to implementing and strengthening the NDCs, to meeting the transparency requirements and to formulating low-emission development plans.

Christiana Figueres, Executive Secretary to the UNFCCC, said recently that getting 196 governments to agree in Paris was the easy part. Moving from words to action, transforming the global economy, rapidly accelerating action: these are the imperatives of the post-Paris world and the hallmarks of the new climate change game that demand concerted action by national and sub-national governments, business, civil society, international organizations and citizens alike. The Paris Agreement provides the directional signal but it is through its implementation, through the impact on the atmosphere and on peoples and economies, that its ultimate success will be judged. And in this endeavour, time is of the essence.

Disclosure statement

No potential conflict of interest was reported by the author.

Notes

1. Officially also the eleventh session of the Conference of the Parties serving as the meeting of the Parties to the Kyoto Protocol.

2. 17,150 Party delegates, 6300 NGO delegates, 1025 delegates from intergovernmental organizations, 900 from United Nations organizations and 2750 press.
3. 78 Presidents and 48 Prime Ministers attended the Leaders Event and delivered speeches and a further 7 Presidents and 7 Prime Ministers attended without speaking. For the full list of speakers, seehttp://newsroom.unfccc.int/cop21parisinformationhub/cop-21cmp-11-information-hub-leaders-and-high-level-segment/.
4. 156 INDCs were received before the start of the Conference from 184 Parties (the European Union submissions cover 28 member states and the EU). Since then, a further xx have been received [to be updated at point of publication – editor].
5. The announcement of albeit modest amounts (in comparison with the negotiations over billions and trillions) for the Adaptation Fund (US$ 75 million) and the Least Developed Countries Fund (US$ 248 million) had significant political impact because these Funds are particularly important for the most vulnerable countries and are relatively easy to access. This contributed to the overall positive mood on finance despite other objectives not being achieved.
6. See for example the NAZCA Portal (http://climateaction.unfccc.int), which registers commitments to climate action by companies, cities, sub-national regions and investors to address climate change.
7. US President Barack Obama's Mission Innovation, Indian Prime Minister Narendra Modi's Solar Alliance, and Bill Gates' Clean Tech Initiative, the Friends of Fossil Fuel Subsidy Reform and the World Bank-International Monetary Fund's global Carbon Pricing initiative are the highest profile examples. See Northrop (2016) for an unofficial listing of related announcements.
8. The Agreement's Preamble "recognize[es] the importance of the engagements of all levels of government and various actors, in accordance with respective national legislations of Parties, in addressing climate change". Otherwise, references to action by "non-Party stakeholders" is relegated to the accompanying decision (UNFCCC 2015a, paras 116–121).
9. Solar energy has dropped 80% in cost since 2008 and the cost of wind has dropped by 40% (IRENA, 2014, pp. 14–15).
10. See NPR (2016) on Morocco's new concentrated solar power plant, set to be the world's largest.
11. The ADP had been the forum for the negotiations leading up to Paris since the mandate that launched the negotiation process at COP 17 in Durban in 2011.

References

BBC. (2015). COP21 climate change summit reaches deal in Paris. *BBC*. Retrieved from http://www.bbc.co.uk/news/science-environment-35084374

Bloomberg. (2016). *Clean energy investment by the numbers, end of year 2015*. Retrieved from http://www.bloomberg.com/company/clean-energy-investment/

DivestInvest. (2016). *DivestInvest Philanthropy report*. Retrieved from http://divestinvest.org/philanthropy/wp-content/uploads/sites/3/2016/01/12.22_Divest-Invest-Brief_web2.pdf

Hood, M. (2016, January 18). 2015 a 'tipping point' for climate change: Experts. *Phys.org*. Retrieved from http://phys.org/news/2016-01-climate-experts.html

IRENA. (2014). *Rethinking energy*. Retrieved from http://www.irena.org/rethinking/Rethinking_FullReport_web.pdf

New York Times. (2015). Nations approve landmark climate accord in Paris. *New York Times*. Retrieved from http://www.nytimes.com/2015/12/13/world/europe/climate-change-accord-paris.html?_r=0

Northrop, M. (2016, January 11). Is the post-fossil fuel era now inevitable? *Huffpost Green*. Retrieved from http://www.huffingtonpost.com/michael-northrop/is-the-post-fossil-fuel-e_b_8954310.html?utm_hp_ref=climate-change

NPR. (2016, February 4). Morocco unveils a massive solar power plant in the Sahara. *NPR*. Retrieved from http://www.npr.org/sections/thetwo-way/2016/02/04/465568055/morocco-unveils-a-massive-solar-power-plant-in-the-sahara

UNEP. (2015). *The Emissions Gap Report 2015*. Nairobi: Author.

UNFCCC. (2015a). *Decision 1/CP.21, in report of the conference of the parties on its twenty-first session, held in Paris from 30 November to 13 December 2015. Addendum Part two: Action taken by the Conference of the Parties at its twenty-first session* (FCCC/CP/2015/10/Add.1).

UNFCCC. (2015b). *Synthesis report on the aggregate effect of the intended nationally determined contributions* (FCCC/CP/2015/7).

UNFCCC news. (2015a). France to invest 2 Billion Euros in Africa Renewables: President Hollande announces commitment at COP21. *UNFCCC News*. Retrieved from http://newsroom.unfccc.int/financial-flows/france-to-invest-2-billion-euros-in-renewables-in-africa/

UNFCCC news. (2015b). India and France launch International Solar Energy Alliance at COP21. *UNFCCC News*. Retrieved from http://newsroom.unfccc.int/clean-energy/international-solar-energy-alliance-launched-at-cop21/

UNFCCC news. (2016). *List of recent climate funding announcements: Updating in Interactive Graphic*. Retrieved from http://newsroom.unfccc.int/financial-flows/list-of-recent-climate-funding-announcements

UN News Centre. (2015). COP21: UN chief hails new climate change agreement as 'monumental triumph'. UN News Centre.

The Paris Agreement: resolving the inconsistency between global goals and national contributions

NIKLAS HÖHNE, TAKESHI KURAMOCHI, CARSTEN WARNECKE, FRAUKE RÖSER, HANNA FEKETE, MARKUS HAGEMANN, THOMAS DAY, RITIKA TEWARI, MARIE KURDZIEL, SEBASTIAN STERL, AND SOFIA GONZALES

The adoption of the Paris Agreement in December 2015 moved the world a step closer to avoiding dangerous climate change. The aggregated individual intended nationally determined contributions (INDCs) are not yet sufficient to be consistent with the long-term goals of the agreement of 'holding the increase in global average temperature to well below 2°C' and 'pursuing efforts' towards 1.5°C. However, the Paris Agreement gives hope that this inconsistency can be resolved. We find that many of the contributions are conservative and in some cases may be overachieved. We also find that the preparation of the INDCs has advanced national climate policy-making, notably in developing countries. Moreover, provisions in the Paris Agreement require countries to regularly review, update and strengthen these actions. In addition, the significant number of non-state actions launched in recent years is not yet adequately captured in the INDCs. Finally, we discuss decarbonization, which has happened faster in some sectors than expected, giving hope that such a transition can also be accomplished in other sectors. Taken together, there is reason to be optimistic that eventually national action to reduce emissions will be more consistent with the agreed global temperature limits.

Policy relevance

The next step for the global response to climate change is not only implementation, but also strengthening, of the Paris Agreement. To this end, national governments must formulate and implement policies to meet their INDC pledges, and at the same time consider how to raise their level of ambition. For many developing countries, implementation and tougher targets will require financial, technological and other forms of support. The findings of this article are highly relevant for both national governments and support organizations in helping them to set their implementation priorities. Its findings also put existing INDCs in the context of the Paris Agreement's global goals, indicating the extent to which current national commitments need to be strengthened, and possible ways in which this could be done.

1. Introduction

The Paris Agreement, adopted at the twenty-first session of the Conference of the Parties (COP 21) to the United Nations Framework Convention on Climate Change (UNFCCC) in December 2015 (UNFCCC, 2015a), set a major landmark in the history of international climate policy. The Paris

Agreement not only stipulates the goal of 'holding the increase in global average temperature to well below 2°C above pre-industrial levels' and 'pursuing efforts' to limit it to 1.5°C (UNFCCC, 2015a, Article 2), but also a corresponding long-term GHG emissions reduction goal of achieving 'balance between anthropogenic emissions by sources and removals by sinks' of GHGs in the second half of the twenty-first century (UNFCCC, 2015a, Article 4.1). Almost 190 countries have submitted intended nationally determined contributions (INDCs), which will become an integral part of the agreement. The Paris Agreement also includes many possible steps to raise ambition along the way. A global stocktake has been agreed to regularly monitor the progress towards the long-term goals every five years, starting with a preparatory version in 2018.

The translation of actions to national and global GHG emissions, and then to temperature increase and damages, still involves a number of uncertainties (IPCC, 2014). It becomes clear (and is now also recognized in the Paris Agreement), however, that the cumulative CO_2 emissions largely determine the resulting temperature increase and hence global CO_2 emissions need to reach net zero at one point.

The process for successful implementation of the Paris Agreement has just begun, and what really matters is the action that will take place in the years to come. To date, a wide spectrum of commentaries on the Paris Agreement have been published worldwide,[1] including a number of articles in the scientific literature on the role of research and scientists in limiting global warming to within 1.5°C (Hulme, 2016; Peters, 2016; Rogelj & Knutti, 2016). However, there are few studies that discuss the policy implications and necessary actions for effective implementation of the Paris Agreement by reviewing the progress in climate policy action and the policy-making process observed at both national and multilateral levels in recent years, in particular in the lead-up to COP 21.

In order to identify key thematic areas and elements for enhanced action and ambition, it is useful to conduct a short review on: (1) the progress made by the Paris Agreement and its lead-up process in light of the ultimate objective of the UNFCCC;[2] (2) the scale of the gap between current mitigation actions committed to by national governments and the action levels required to meet the Paris Agreement's long-term goals; and (3) the mechanisms under the Paris Agreement to ratchet up ambition levels and the challenges that nations will face in doing so.

This article is partly based on a number of recent findings by the authors (in particular sections 3 and 4.2). It focuses exclusively on national and global mitigation goals and actions in relation to the Paris Agreement. The article does not discuss other issues, such as adaptation to climate change, the provision of climate finance or loss and damage (on these issues, see Sharma, 2016 and Lees, 2016).

This article is structured as follows. In Section 2 we review the outcomes of COP 21 in comparison with earlier expectations. Second, we discuss the level of current ambition by comparing the aggregate impact of INDCs with the emission levels required to achieve the long-term goals stipulated in the Paris Agreement (Section 3). Third, we review how the international process in the lead-up to COP 21 contributed to the development of processes and institutions for national climate policy-making (Section 4). Following this, this article discusses how the ratchet mechanism of continually raising ambition towards the 1.5°C or 2°C goal can be implemented in an effective manner, and how non-state mitigation actions can be promoted and implemented to maximize their significant potential impact (Section 5). Finally, in Section 6, we make a number of general recommendations for the successful implementation of the Paris Agreement.

2. The Paris COP: expectations and outcomes

The outcomes of the Paris COP surpassed the expectations of many. This may be partly because the mismatch between the expectations and results of the last major attempt to agree on a global treaty on climate change (Bailey, 2010), in Copenhagen in 2009, may have led to relatively low expectations before the Paris conference.

Research organizations had prepared draft text proposals for a possible Paris Agreement (e.g. Oberthür, La Vina, & Morgan, 2015; Yamin, Haites, & Höhne, 2015), which were surpassed in many areas by the final agreed text. The Paris outcomes were indeed generally welcomed by environmental NGOs and think tanks (Greenpeace, 2015; WRI, 2015). In the authors' view, several newly agreed elements were remarkable.

First, the fact that almost all countries submitted their INDCs before Paris (160 submissions representing 188 countries of the 196 Parties to the UNFCCC) (UNFCCC, 2016b) was an unprecedented engagement that had not been predicted. For example, based on a country survey in March 2015, we estimated that around 80 INDCs would have been submitted before COP21 (NewClimate, 2015). It turned out to be more than double.

Second, the firm inclusion of the clause on 'pursuing efforts to limit the temperature increase to 1.5°C above pre-industrial levels' came as a surprise to many (UNFCCC, 2015a, Article 2). Before the conference, even the 2°C target was under huge pressure (Geden, 2015), being called unrealistic, but the Paris Agreement moved the goal to an even more ambitious level. The agreement on the long-term net zero emission goal was also very uncertain before Paris. It had been framed as a global goal initially by researchers (Haites, Yamin, & Höhne, 2013) and had been supported later by politically prominent actors including the Organisation for Economic Co-operation and Development (OECD) Secretary-General in 2013 (Gurría, 2015) and the G7 leaders (G7, 2015). An expert survey conducted to make a forecast of key outcomes from COP21 showed that more than 75% of respondents predicted that there would be either no goal or only a qualitative target for 2050, and less than a third predicted there would be a quantifiable goal of net zero emissions or net negative emissions, even by 2100 (Kallbekken & Sælen, 2015).

Third, the Paris Agreement includes many possible steps to raise ambition along the way, which is an implicit recognition that the current proposals by countries are not yet sufficient in aggregate. The global stocktake to regularly monitor progress towards the long-term goals is also an important component. Although the details will be determined in future meetings, it is worth noting that the process to raise ambition enshrined in the Paris Agreement is very similar to what was proposed earlier by a group of climate policy experts calling for high ambition (Morgan, Dagnet, Höhne, & Oberthür, 2014) (see Section 5 for details).

One of the many factors that contributed to the positive outcomes of COP 21 was the increasing number of calls for strengthened climate action from sectors other than academia and environmental NGOs. For example, a number of studies conducted by the New Climate Economy (NCE), the commission of which comprised former heads of government and finance ministers as well as leaders in economics, business and finance, emphasized that lasting economic growth and GHG emissions reductions could be achieved at the same time (NCE, 2014, 2015). Moreover, many business-based and subnational government-based initiatives for enhanced climate action emerged in the lead-up to Paris (discussed in detail in Section 4.4).

3. The current level of national mitigation ambition in comparison with the long-term goals of the Paris Agreement

While it is welcomed that the Paris Agreement and its preparation process raised the global long-term goal for climate change mitigation, the current level of national action under the Paris Agreement is still not in line with the agreed global long-term goals.

In this section, we briefly describe the results from the Climate Action Tracker (CAT) project (http://www.climateactiontracker.org). We are main contributors to the CAT, which analysed the emissions trajectories under current policies and the potential impact of the INDCs for 30 countries with high emissions (Jeffery et al., 2015). The basis for the analysis was national studies, complemented with estimates of the latest implemented policies. For the quantification of the INDCs we based the estimates on data provided by the countries, scrutinized and complemented by external sources. We aggregated all of the submitted INDCs (at the time of publication), calculated the expected future GHG emission levels and calculated the global temperature increase using the simple carbon cycle climate model MAGICC (Hulme et al., 2000; Meinshausen, Raper, & Wigley, 2011; Meinshausen, Wigley, & Raper, 2011) in a probabilistic mode. More information on the method is provided in the Supplemental Data. The results are presented in Figure 1.

It was found that, without any climate policy, GHG emissions would increase further during the course of the century (the baseline in Figure 1). With currently implemented climate policies in all

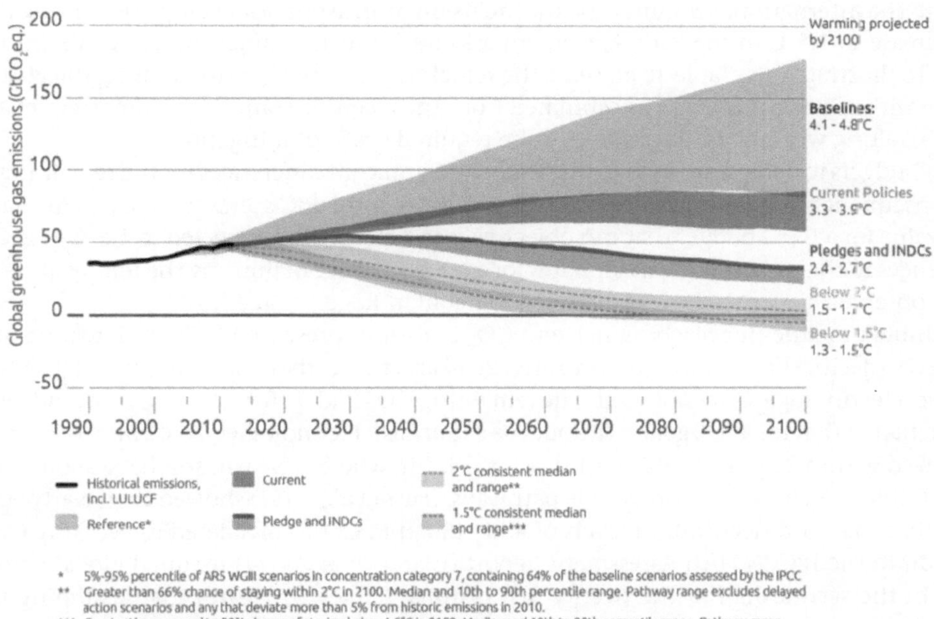

Figure 1. Global GHG emissions under different scenarios and the resulting temperature increase by 2100.
Source: (Jeffery et al., 2015).

countries, global GHG emissions would stabilize in the second half of the twenty-first century and lead to a temperature increase of around 3.6°C by 2100 (with a range of 2.2–3.9°C) with an increasing trend thereafter (labelled 'current policies' in Figure 1). With full implementation of the INDCs declared by all countries, global emissions could start to decline in the second half of the century and lead to a temperature increase of 2.7°C (with a range of 2.2–3.4°C) by 2100 (increasing thereafter) or lower, if the conditional INDCs are also taken into account (labelled 'pledges and INDCs' in Figure 1). However, this path is still far away from what would be necessary to limit temperature increase below 2°C or 1.5°C (the lowest scenarios in Figure 1).

The INDCs, as declared in the lead-up to and during the COP 21, therefore bring us roughly one third of the way in 2030 towards a pathway that would be compatible with keeping the global average temperature increase well below 2°C.

The GHG emissions projections of the CAT are consistent with those from other similar models, including nine studies covered alongside the CAT in the 2015 Emissions Gap Report of the UN Environment Programme (UNEP, 2015), as well as with those from the UNFCCC report on the aggregation of INDCs (UNFCCC, 2015b). With regard to the temperature increase projections, it should be noted that other studies project even higher temperature increases (median estimate), up to 3.7°C (Levin & Fransen, 2015). The finding that INDCs are not yet consistent with the long-term goals of the Paris Agreement is robust even under different interpretations of these long-term goals, which do not specify the timing when, or probability with which, 1.5°C or 2°C should be reached.

Although limiting the global average temperature increase to within 2°C already poses a significant challenge to the international community, the inclusion of an aspirational target of limiting the temperature increase to 1.5°C in the Paris Agreement asks for mitigation ambition to be raised to a whole new level. To illustrate this, Table 1[3] presents the remaining CO_2 budgets for keeping the global temperature rise within 1.5°C and 2°C [4] as estimated from the recent literature (Friedlingstein et al., 2014; IPCC, 2014), along with the implications for the required speed of mitigation.

The CO_2 budgets in Table 1 are used to illustrate the significant differences for the remaining budgets and the consequent mitigation pathways between the 2°C and 1.5°C scenarios, cognizant that there is no single value for a GHG budget that can objectively and clearly be connected to 1.5°C or 2°C targets. The differences in the carbon budget estimates for a >66% chance of limiting the temperature increase below 2°C observed across studies are analysed in detail in Rogelj et al., 2016.

The example timelines for reaching net zero CO_2 emissions presented in Table 1 assume that emissions started to decline linearly towards zero after 2014 and that carbon dioxide removal (CDR) or negative-emission technologies[5] are not used. The remaining CO_2 budget for a 1.5°C goal would be reduced to less than half of that for a 2°C goal. Without the use of CDR technologies, zero emissions would need to be achieved within 20 years to stay within a 1.5°C limit, whereas we will still have about 40 years in case of a 2°C limit. Even for 2°C-compatible pathways, (Fuss et al., 2014) showed that nearly 90% of the pathways that lead to concentration levels of 430–480 ppm CO_2e considered in Working Group III's contribution to the IPCC's Fifth Assessment Report (Clarke et al., 2014) required global net negative emissions in the second half of the twenty-first century. Table 1 indicates that, assuming no large-scale deployment of CDR technologies, in particular bioenergy combined with carbon capture and storage (BECCS), in the mid- to long-term future, the 1.5°C limit is only achievable with a complete phase-out of existing fossil fuel-based infrastructure within less than 20 years of now. This would

Table 1 Remaining CO_2 budgets for keeping the temperature within 1.5 °C and 2 °C and the example timelines for reaching net zero CO_2 emissions under the assumptions that the emissions started to decline linearly towards zero after 2014 and CDR technologies as defined by the IPCC (2014) are not used

Indicator		Units	Global average temperature increase above pre-industrial levels: 2 °C (fraction of simulations that meet the goal: 66%)	1.5 °C (fraction of simulations that meet the goal: 50%)
Remaining CO_2 budget[1]	end-2010	$GtCO_2$	1000	550
	end-2014	$GtCO_2$	840	390
The year net zero CO_2 emissions is achieved under the assumptions given[2]		Year	2056	2034

[1]The values for remaining CO_2 budget at the end of 2010 are based on the results for Representative Concentration Pathways from complex models presented in the IPCC AR5 Synthesis Report (IPCC, 2014). The cumulative CO_2 emissions for 2011–2014 were estimated to be 157 $GtCO_2$ based on Friedlingstein et al. (2014). The remaining budgets at the end of 2014 are rounded to 10 $GtCO_2$.
[2]Total CO_2 emissions in 2014 are estimated to be 40.3 $GtCO_2$ (Friedlingstein et al., 2014).

require shutting down a significant amount of fossil fuel-based installations before the end of their life with significant associated costs.

4. Potential strengthening of INDCs

This section discusses several factors that could lead to a strengthening and/or overachievement of INDCs, so that eventually national action to reduce emissions could be made more consistent with the agreed global long-term goals.

4.1. Conservative nature of INDCs

Some countries have understandably been cautious when submitting their INDCs. This may be partly due to uncertainty over the rules that would be adopted under the Paris Agreement, and in particular the extent to which countries would be legally bound to their contributions. Other countries may have had other reasons to hesitate to make any strong substantive commitments, e.g. concerns over equity, economic implications, possibility of 'losing face' by failing to meet targets, lack of experience in setting targets and concerns that the promised support in terms of finance, technology and capacity building would not be forthcoming.

There are indications that several countries considered varying degrees of ambition, but eventually chose not to put forward the most ambitious one. Examples include Vietnam (initial target of 12.5% below business as usual (Department of Meteorology, Hydrology and Climate Change of Vietnam, 2015) in 2030 compared with the final INDC of 8%) and the EU (the impact assessment (EC, 2014) that formed the basis for the decision included several target options, including a reduction of 45% below 1990 in 2030, compared with 'at least 40%' in the INDC). These are largely political decisions to minimize the risk of governments failing to achieve what they pledged, and they imply that further mitigation can be achieved in these countries in a technically and economically reasonable manner. In the case of Japan, its INDC became less ambitious partly due to the assumption of very optimistic future GDP growth rates, which increased the baseline emission levels to be consistent with economic policy (Government of Japan, 2015; Kuramochi, 2015; Kuramochi, Wakiyama, & Kuriyama, 2015).

For some countries, their INDCs do not constitute a great deviation from the GHG emissions projections under currently implemented policies (CAT, 2015), e.g. Argentina (Climate Action Tracker, 2015a) Brazil (Climate Action Tracker, 2015b) China (Climate Action Tracker, 2015c) EU (Climate Action Tracker, 2015d) India (Climate Action Tracker, 2015e) Indonesia (Climate Action Tracker, 2015f) and Japan (Climate Action Tracker, 2015g). This implies that these countries would not have to strengthen their mitigation policies from current levels to meet their INDCs and that they could overachieve their INDCs by implementing additional policies between now and 2025 or 2030. For other countries, like Russia or Ukraine, meeting the INDCs would result in even higher emissions than current trends (Jeffery et al., 2015). They, in particular, could revisit their INDCs.

Finally, restructuring and trends change quite quickly in some countries, which may not yet be fully represented in the INDCs. For example, China's economic restructuring may lead to CO_2 emissions from energy growing at a much slower rate than under the old economic model (den Elzen et al., 2016; Green & Stern, 2016).

The Paris Agreement is now adopted with almost universal support. It defines the legal nature of the contributions and clarifies that punitive measures for non-compliance with the proposed contributions are absent. Although there are various reasons that preclude national governments from committing to ambitious mitigation targets as indicated above, the flexible nature of the Paris Agreement could potentially encourage countries to revisit their proposals and possibly strengthen them (see Section 4.3).

It should also be noted that, for a smaller number of countries, the INDCs imply a significant deviation from current trends and require significant additional national polices to achieve this change, such as Australia, Canada and the US (Jeffery et al., 2015). Nevertheless, this alone does not mean that the INDCs are ambitious; the INDCs of these three countries mentioned here are evaluated by the CAT to be 'inadequate' (Climate Action Tracker, 2015g, 2015h, 2015i, 2015j)

4.2. Advancement of national climate policy-making through the preparation for Paris

A second reason for hope is that the preparation of the INDCs for the new international climate agreement under negotiation had advanced national policy-making even before the Paris Agreement was adopted. The international pressure to produce an INDC accelerated national climate policy-making processes significantly.

In October 2015, we carried out a survey among government representatives, national consultants and supporting institutions involved in the national INDC preparation processes, see Figure 2 (Day, Röser, Tewari, Kurdziel, & Höhne, 2015). We obtained information from stakeholders in 52 countries, covering all major regions. The results can only give an indication as they are from a non-representative sample that includes mainly those heavily involved in the preparation process. However, several indications given by these respondents were very clear.

First, the preparation of INDCs kick-started climate planning and strategy development processes and consolidated and built upon existing climate strategy and planning processes, as confirmed by

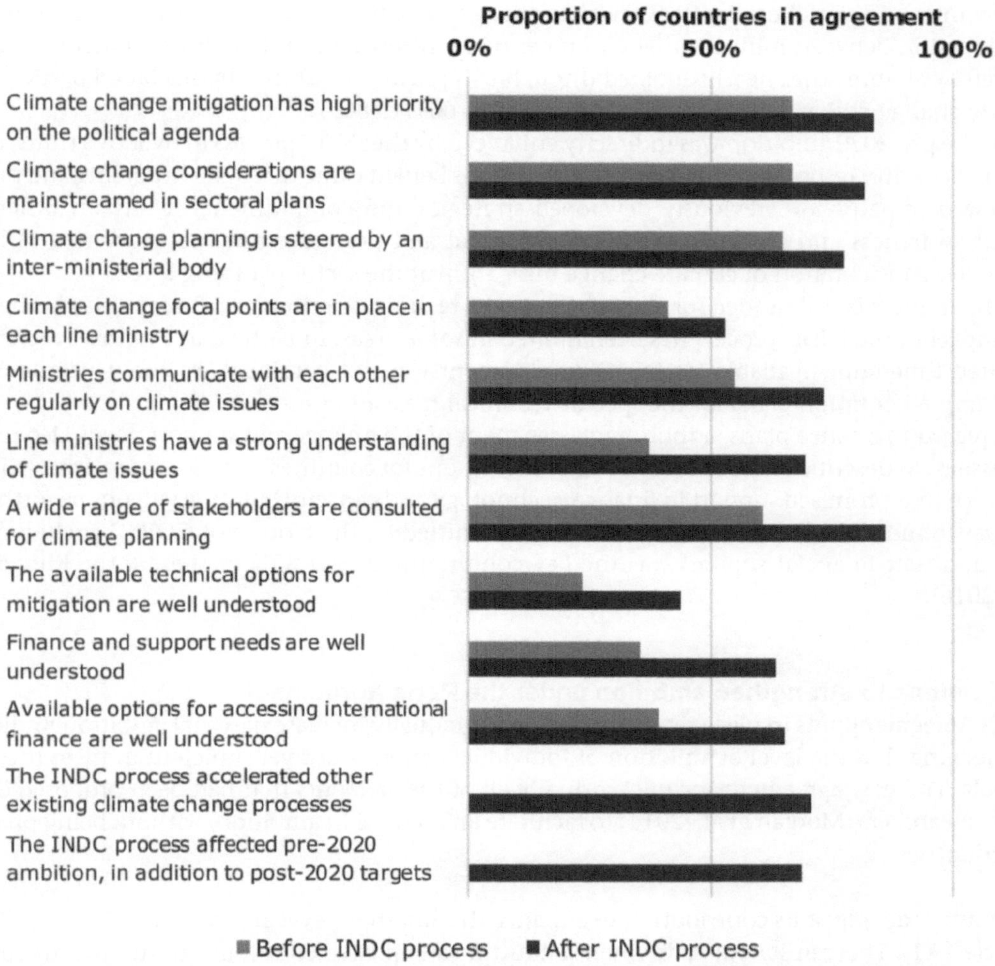

Figure 2. Opinions on the benefits from the process of preparing the INDCS from a survey of stakeholders from 52 countries.
Source: (Day et al., 2015).

over 70% of the respondents from developing countries. The concentrated international focus brought on by the INDC process catalysed new or renewed momentum at the country level for the development of new policies and legislation and new action plans for implementation that reach down to the sectoral level.

Second, the INDC process significantly advanced national political agendas. Climate change mitigation is now a high political priority for the vast majority of consulted developing countries (approximately 84%, compared with 67% before), which may represent a new critical mass for enhanced international cooperation and negotiation. Institutional structures were developed or strengthened, with new interministerial committees, climate change focal points in line ministries and enhanced communication between ministries. The number of countries stating that climate change is understood well by all ministries nearly doubled through this process. Countries with a broad participation in climate change policy-making have increased (from 60% to 80%).

Third, the pre-2020 ambition was indirectly enhanced by the INDC process in over two thirds of the developing country respondents. In some countries, this benefit came through reinforcing and providing renewed impetus for previously developed strategies and commitments, whilst in others the increased awareness and capacity across governmental and non-governmental stakeholders led to more regular consideration of climate change mitigation at the sector planning level.

Fourth, progress is still needed for elaboration of the technical options and finance plans. The catalysing impact of the INDC process has been limited in some areas. In particular, countries report that the limited timeframe available for INDC development was not conducive to the development of detailed implementation plans for the specific technical measures proposed.

With regard to finance plans, a thorough assessment of the finance and support needs also remain critical issues. As described in Section 4.1, one of the reasons for countries setting conservative INDCs is the concern that promised support in finance, technology and capacity building will not be forthcoming. It was found that at least 78% of the INDCs submitted by the closure of COP 21 included conditions, and that financial support was the key condition for over 80% of these (Day, Kurdziel, & Roeser, 2016).

4.3. Provisions to strengthen ambition under the Paris Agreement

The Paris Agreement puts in place several elements to gradually increase national ambition, indirectly acknowledging that the level of ambition of individual action is not yet sufficient to meet the long-term goals. The new agreement includes virtually all of the elements that had been proposed earlier by some researchers (Morgan et al., 2014) to facilitate an increase in ambition, without being prescriptive to countries:

- A country can adjust its contribution to enhance the ambition level at any time (UNFCCC, 2015a, Article 4.11). This can start already when submitting an updated contribution with the instrument of ratification of the agreement (UNFCCC, 2015a, Decision paragraph 22).
- Governments will need to update their actions every five years and each successive update has to be at least as strong as the current one (UNFCCC, 2015a, Article 4.3). The new or renewed contributions for 2030 have to be submitted 9–12 months in advance of the COP in 2020 (UNFCCC, 2015a, Decision paragraph 23/24).
- Each country's actions will be reviewed and individual suggestions for improvement for each country will be provided during that review.[6]

Tracking the progress of individual and collective contributions is another important achievement of the Paris meeting. With regards to the global stocktake, a facilitative dialogue on stocktaking will occur in 2018 and the first full global stocktake will take place in 2023 (UNFCCC, 2015a, Decision paragraph 20 and Article 14). This exercise will occur every five years thereafter. To assist countries in the implementation of their actions, a facilitative implementation committee of experts has been agreed (UNFCCC, 2015a, Article 15). The exact modalities of the review remain to be determined in the years to come (Röser, Fekete, Höhne, & Kuramochi, 2015).

The Paris Agreement also increased the need for long-term management of cumulative GHG emissions at a national level to ensure consistency with long-term goals. Simply setting a 2050 target, for instance, is no longer sufficient, and regular carbon management will be needed – as implemented, for example, in the UK under the Climate Change Act (UK Government, 2008). The assessment of the remaining carbon budgets at a national level consistent with the 1.5°C or 2°C limit, and the corresponding emission pathways towards net zero emissions – as carried out in Fekete, Hagemann, & Höhne, 2013; Kuramochi, Asuka, Fekete, Tamura, & Höhne, 2015, among others – may become increasingly important for national climate policy-making in the coming years.

The detailed rules of the ambition cycle of the Paris Agreement, and its implementation, will determine if it can effectively raise ambition. The limited carbon budgets for 2°C, and in particular for 1.5°C, suggest that existing contributions for 2020, 2025 and 2030 have to be strengthened; if these goals are to be reached, it will be too late to only submit strengthened contributions for new and later years. Ambition must already be increased up to 2020. This will only work if the Paris Agreement is ratified quickly alongside the prompt development of the rules for the global stocktake and the review procedures.

4.4. Non-state action

Another of the positive elements of the Paris conference was the unprecedented number of actions and announcements of so called non-state actors or non-Party actors such as cities, regions, companies and sectors. Non-state actions cover a wide range of activities; some of these are international, which are often referred to as international cooperative initiatives (ICIs), while others are domestic. The COP decision adopting the Paris Agreement (UNFCCC, 2015a, Decision paragraphs 118–120 and 134–137) acknowledges the importance of non-state actors and encourages them to undertake enhanced mitigation actions. As of July 2016, more than 11,500 non-state actions and initiatives had been registered under the UNFCCC portal (UNFCCC, 2016a). It is expected that more international non-state actions will be declared and implemented following the Paris Agreement.

Although there are only a few studies available in the literature, they indicate that the combined potential impact of mitigation initiatives by non-state actors could make a significant contribution to achieving the 2°C or 1.5°C goal. The 2015 UNEP Emissions Gap Report reviewed and compared four recent studies on ICIs, all of which were prepared well before the extensive tabling of INDCs in the run-up to Paris. It concluded that the potential global impact of existing ICIs could already be as large as 0.75–2.2 $GtCO_2e$ by 2020, after taking into account overlaps among initiatives as well as any overlap with the national pledges (UNEP, 2015). The first available grey literature that includes the announcements of Paris and extends the analysis to 2030 (Graichen et al., 2016) estimates that, if the 19 initiatives analysed reach their stated goals, they together would potentially deliver 6–11 $GtCO_2e$ of additional emission reductions compared with the INDC emission pathway in 2030. Peer reviewed literature, however, is not yet available.

National governments are probably not yet capturing the entirety of the mitigation commitments made by their subnational governments and businesses in their INDCs. A case study on Germany, for example, suggests that a selected set of quantifiable commitments that have already been made by non-state actors would reduce emissions by an amount equivalent to 0.8–1.7% of 1990 emissions, in addition to the implemented national policies, by 2020 (Höhne, Sterl, & Fekete, 2015). At the global level, some targets of ICIs, e.g. the Climate and Clean Air Coalition (CCAC) on the reduction of non-CO_2 GHGs, are more ambitious than national-level policies planned under the INDCs (Graichen et al., 2016).

It should be noted, however, that the mitigation impact of non-state actions requires that the non-state actors live up to their commitments (Graichen et al., 2016; UNEP, 2015). Non-state actions are, by definition, voluntary and may not have a robust system to measure, report and verify (MRV) their mitigation impacts. Enhanced transparency would therefore be key to identifying real progress in mitigation delivered by non-state actions (IVM, 2015).

4.5. Technological breakthroughs

Finally, it is encouraging that some sectors are experiencing a faster low-carbon transformation than expected. The recognition of the significant co-benefits of such a transformation (NCE, 2014) may have supported this trend.

For example, the development of some renewable energy technologies (such as solar photovoltaics and onshore wind) has consistently been faster than projected (Cronin et al., 2015; Mai et al., 2013; Metayer, Breyer, & Fell, 2015), although still short of achieving scenarios consistent with the 2°C goal (IEA, 2015). The policy interventions of a small group of countries, including Denmark and Germany, has supported the fast uptake of renewables and stimulated mass production and further support policies, particularly in China. The associated cost reductions have made electricity generation from wind and solar economically competitive in many regions of the world.

Most recently the use of electric cars has seen an unprecedented boost. Again, a small set of countries and regions, in this case China, Norway and California, has supported the deployment of the technology to such an extent that the cost reductions have had global repercussions. In Norway, which currently has the highest rate of electric cars per capita globally, 20% of newly registered cars are electric (ICCT Europe, 2014). Sales have doubled annually between 2012 and 2015 (Aasness & Odeck, 2015).

At the same time, there are countervailing trends, such as the discovery of new fossil fuel resources, increased transport demand and faster-than-expected economic development in some regions. Moreover, as described in Section 3, staying within the 1.5°C limit will in the long term require the massive deployment of BECCS and other CDR technologies. However, CDR technologies are still underdeveloped, emerging more slowly than expected,[7] and are associated with large challenges, risks and uncertainties, which are described in detail in the WGIII Contribution to IPCC AR5 (Bruckner et al., 2014; Clarke et al., 2014). It is warned that insufficient near-term mitigation efforts would result in a heavy reliance on CDR technologies such as BECCS, and put future decision makers under greater pressure (Clarke et al., 2014). Policy makers around the world would be required to carefully develop a long-term strategy towards the very large-scale deployment of CDR technologies.

Nevertheless, the positive developments observed in recent years provide evidence that a low-carbon transition that is faster than expected is possible, and that such change can be influenced by the determined actions of a rather small number of countries. The massive scaling up of such activities will be necessary to achieve the long-term goals of the Paris Agreement, acknowledging that some will inevitably fail.

5. Conclusions

In this article we reviewed the status of mitigation actions after the Paris Agreement and confirm that the pledged national action is not compatible with the agreed long-term goals of the Paris Agreement. But we also identified a number of elements that point to avenues for raising global ambition to the level consistent with the Paris goals. We conclude that to this end three major steps are needed.

First, the INDCs need to be rapidly and fully implemented; in most cases they need to be, and can be, overachieved. The momentum created in Paris must be captured by building on the positive experiences with INDC planning, and continuing with the same processes to turn them towards effective implementation. These processes need to include efforts to scale up action immediately. If mitigation action does not scale up significantly before the preparatory global stocktake in 2018, or even up to 2020, it will be very difficult to make up for the time lost.

Second, the leading non-state actors need to be supported, and their efforts need to be captured in national ambition. After the unprecedented involvement of non-state actors in Paris, an increasing number of such actors are likely to set mitigation goals and implement mitigation actions that go beyond the ambition level of their respective national governments. These actions should be supported both financially and institutionally so that they actually deliver substantial emissions reductions that can be monitored and verified.

Finally, the faster-than-expected transformations in some sectors can be used as a model by new 'transformative coalitions': groups of countries and other stakeholders of sufficient critical mass that have the intention to flip global markets by deploying new technologies at a large scale. They would support market uptake, and with that technology development on a sufficient scale to transform not only their domestic markets, but also the global market. Potential areas include zero-energy buildings, efficient electrical appliances, electricity storage, zero-emissions aviation and zero-emissions cement or steel.

Taken together, there is reason to be optimistic that national action to reduce GHG emissions will eventually be more consistent with the agreed global temperature limits of the Paris Agreement. The window of opportunity for reaching 2°C, let alone 1.5°C, without massive deployment of CDR technologies, is closing rapidly. A reinforcing upward spiral of national government policy, non-state actions and transformative coalitions will be essential if dangerous climate change is to be avoided.

Acknowledgements

The calculations of the CAT were performed by the CAT team consisting of Louise Jeffery, Claire Fyson, Ryan Alexander, Johannes Gütschow (PIK), Marcia Rocha, Jasmin Cantzler, Michiel Schaeffer, Bill Hare (Climate Analytics), Markus Hagemann, Hanna Fekete, Sebastian Sterl, Takeshi Kuramochi, Niklas Höhne (NewClimate Institute), Pieter van Breevoort, Yvonne Deng, Karlien Wouters and Kornelis Blok (Ecofys). The survey on INDC preparation was performed by Thomas Day, Frauke Röser, Ritika Tewari, Marie Kurdziel and Niklas Höhne (NewClimate Institute).

Disclosure statement

No potential conflict of interest was reported by the authors.

Supplemental data

Supplemental data for this article can be accessed at 10.1080/14693062.2016.1218320.

Notes

1. A non-exhaustive list of commentaries can be found ina blog post of the Global Environmental Policy Program of the School of International Service, American University: http://teachingclimatelaw.org/compendium-of-commentary-on-the-paris-agreementcop21 (accessed 9 March, 2016).
2. The ultimate objective of the UNFCCC is to stabilize GHG concentrations in the atmosphere 'at a level that would prevent dangerous anthropogenic interference with the climate system' (UNFCCC, 1992, Article 2).
3. Inspired by http://www.carbonbrief.org/six-years-worth-of-current-emissions-would-blow-the-carbon-budget-for-1-5-degrees
4. For the 1.5°C, threshold values at which 50% of the simulations from the research projects reviewed in Collins et al. (2013) met the 1.5°C limit. Similarly for 2°C, threshold values at which 66% of the simulations from the research projects reviewed in Collins et al. (2013) met the 2°C limit. There is no data available for >66% for 1.5°C due to the limited number of scenarios. Nevertheless, the combination of >66% for 2°C and >50% for 1.5°C used here adequately represents the goal of the Paris Agreement.
5. CDR methods

   > refer to a set of techniques that aim to remove CO_2 directly from the atmosphere by either (1) increasing natural sinks for carbon or (2) using chemical engineering to remove the CO_2, with the intent of reducing the atmospheric CO_2 concentration (IPCC, 2013, p. 1449).

   CDR methods include large-scale afforestation, BECCS and direct air capture and sequestration (Rogelj & Knutti, 2016).
6. Article 13.12. The relevant text reads

   > The review shall also identify areas of improvement for the Party, and include a review of the consistency of the information with the modalities, procedures and guidelines referred to in paragraph 13 of this Article, taking into account the flexibility accorded to the Party under paragraph 2 of this Article.

   It is ambiguous if the 'areas for improvement' relate only to reporting or to improving policy implementation (UNFCCC, 2015a).
7. As of 2012, there are only two small-scale examples of commercial precursors to BECCS that capture CO_2 emissions from ethanol production facilities for enhanced oil recovery in close-proximity facilities (Bruckner et al., 2014, based on DiPietro, Balash, & Wallace, 2012).

ORCiD

Niklas Höhne http://orcid.org/0000-0001-9246-8759
Takeshi Kuramochi http://orcid.org/0000-0002-3976-0133
Carsten Warnecke http://orcid.org/0000-0003-0689-8897

References

Aasness, M. A., & Odeck, J. (2015). The increase of electric vehicle usage in Norway - incentives and adverse effects. *European Transport Research Review, 7*(34). doi:10.1007/s12544-015-0182-4

Bailey, I. (2010). Copenhagen and the new political geographies of climate change. *Political Geography, 29*(3), 127–129. http://doi.org/10.1016/j.polgeo.2010.02.008

Bruckner, T., Bashmakov, I. A., Mulugetta, Y., Chum, H., Navarro, A. de la V., Edmonds, J., … Zhang, X. (2014). *Energy Systems*. (O. Edenhofer, R. Pichs-Madruga, Y. Sokona, E.Farahani, S. Kadner, K. Seyboth, … J. C. Minx, Eds.) *Climate Change 2014: Mitigation of Climate Change. Contribution of Working Group III to the Fifth Assessment Report of the Intergovernmental Panel on Climate Change*. Cambridge, United Kingdom and New York, NY, USA: Cambridge University Press.

CAT. (2015). *Climate action tracker: Country assessments [ICOMM]*. Retrieved May 20, 2016, from http://climateactiontracker.org/countries.html

Clarke, L., Jiang, K., Akimoto, K., Babiker, M., Blanford, G., Fisher-Vanden, K., … van Vuuren, D. P. (2014). *Assessing transformation pathways*. (O. Edenhofer, R. Pichs-Madruga, Y. Sokona, E.Farahani, S. Kadner, K. Seyboth, … J. C. Minx, Eds.) *Climate Change 2014: Mitigation of Climate Change. Contribution of Working Group III to the Fifth Asessment Report of the IPCC*. Cambridge, UK and New York, NY, USA: Cambridge University Press

Climate Action Tracker. (2015a). *Country page Argentina*. Retrieved May 20, 2016, from http://climateactiontracker.org/countries/argentina.html

Climate Action Tracker. (2015b). *Country page Brazil*. Retrieved May 20, 2016, from http://climateactiontracker.org/countries/brazil.html

Climate Action Tracker. (2015c). *Country page China*. Retrieved May 20, 2016, from http://climateactiontracker.org/countries/China.html

Climate Action Tracker. (2015d). *Country page EU*. Retrieved May 20, 2016, from http://climateactiontracker.org/countries/eu.html

Climate Action Tracker. (2015e). *Country page India*. Retrieved May 20, 2016, from http://climateactiontracker.org/countries/india.html

Climate Action Tracker. (2015f). *Country page Indonesia*. Retrieved May 20, 2016, from http://climateactiontracker.org/countries/indonesia.html

Climate Action Tracker. (2015g). *Country page Japan*. Retrieved May 20, 2016, from http://climateactiontracker.org/countries/japan.html

Climate Action Tracker. (2015h). *Country page Australia*. Retrieved May 20, 2016, from http://climateactiontracker.org/countries/australia.html

Climate Action Tracker. (2015i). *Country page Canada*. Retrieved May 20, 2016, from http://climateactiontracker.org/countries/canada.html

Climate Action Tracker. (2015j). *Country page USA*. Retrieved May 20, 2016, from http://climateactiontracker.org/countries/usa.html

Collins, M., Knutti, R., Arblaster, J., Dufresne, J.-L., Fichefet, T., Friedlingstein, P., … Wehner, M. (2013). *Long-term climate change: Projections, commitments and irreversibility*. (T. F. Stocker, D. Qin, G.-K. Plattner, M. Tignor, S. K. Allen, J. Boschung, … P. M. Midgley, Eds.) *Climate Change 2013: The Physical Science Basis. Contribution of Working Group I to the Fifth Assessment Report of the Intergovernmental Panel on Climate Change*. Cambridge: Cambridge University Press.

Cronin, C., Menon, S., Monteith, S., Plechaty, D., Hagemann, M., Höhne, N., … Sferra, F. (2015) *Faster and cleaner – Decarbonisation in the power and transport sectors is surpassing predictions and offering hope for limiting warming to 2°C*. San Francisco, USA. Retrieved from https://newclimateinstitute.files.wordpress.com/2015/12/faster-cleaner-decarbonization-in-the-power-transport-sectors.pdf

Day, T., Kurdziel, M., & Roeser, F. (2016). *Conditionality of intended nationally determined contributions (INDCs)*. Berlin: Mitigation and MRV Partnership. Retrieved from https://mitigationpartnership.net/conditionality-indcs

Day, T., Röser, F., Tewari, R., Kurdziel, M.-J., & Höhne, N. (2015). *Preparation of intended nationally determined contributions (INDCs) as a catalyst for national climate action*. Cologne. Retrieved from http://mitigationpartnership.net/sites/default/files/indc_as_catalyst.pdf

Department of Meteorology, Hydrology and Climate Change of Vietnam. (2015). Explanation of the INDC. Retrieved March 8, 2016, from http://www.dmhcc.gov.vn/tin-tuc/2591/Hoi-thao-tham-van-lan-thu-2-du-thao-bao-cao-INDC-Viet-Nam.html

DiPietro, P., Balash, P., & Wallace, M. (2012). A note on sources of CO_2 supply for enhanced-oil-recovery operations. *SPE Economics & Management*, *4*(2), 69–74.

den Elzen, M., Fekete, H., Höhne, N., Admiraal, A., Forsell, N., Hof, A. F., ... van Soest, H. (2016). Greenhouse gas emissions from current and enhanced policies of China until 2030: Can emissions peak before 2030? *Energy Policy*, *89*, 224–236. doi:10.1016/j.enpol.2015.11.030

European Commission. (2014). Impact assessment accompanying the document – Communication from the commission to the European parliament, the council, the European economic and social committee and the committee of the regions – A policy framework for climate and energy in the period. Brussels: European Commission. Retrieved from http://eur-lex.europa.eu/legal-content/EN/TXT/PDF/?uri=CELEX:52014SC0015&from=EN

Fekete, H., Hagemann, M., & Höhne, N. (2013). *Australia's carbon budget based on global effort sharing. Technical report. Commissioned by WWF Australia*. Cologne: Ecofys. Retrieved from http://awsassets.wwf.org.au/downloads/fs067a_australia_carbon_budget_based_on_global_effort_sharing_24oct13.pdf

Friedlingstein, P., Andrew, R. M., Rogelj, J., Peters, G. P., Canadell, J. G., Knutti, R., ... Le Quéré, C. (2014). Persistent growth of CO_2 emissions and implications for reaching climate targets. *Nature Geoscience*, *7*(10), 709–715. doi:10.1038/ngeo2248

Fuss, S., Canadell, J. G., Peters, G. P., Tavoni, M., Andrew, R. M., Ciais, P., ... Yamagata, Y. (2014). Betting on negative emissions. *Nature Climate Change*, *4*(10), 850–853. doi:10.1038/nclimate2392

G7. (2015). *Leaders' declaration. G7 Summit*, *7*–8 June 2015. Elmau. Retrieved from https://www.g7germany.de/Content/EN/_Anlagen/G7/2015-06-08-g7-abschluss-eng_en.pdf?__blob=publicationFile&v=3

Geden, O. (2015). Policy: Climate advisers must maintain integrity. *Nature*, *521*(7550), 27–28. doi:10.1038/521027a

Government of Japan. (2015, 17 July). *Submission of Japan's Intended Nationally Determined Contribution (INDC)*. Bonn: United Nations Framework Convention on Climate Change. Retrieved from http://www4.unfccc.int/submissions/indc/published%20documents/japan/1/2015017_japan's%20indc.pdf

Graichen, J., Healy, S., Siemons, A., Höhne, N., Kuramochi, T., Gonzales-zuñiga, S., ... Wachsmuth, J. (2016). *Climate initiatives, national contributions and the Paris Agreement*. Berlin: Oeko Institute. Retrieved from https://newclimate.org/2016/05/23/climate-initiatives-national-contributions-and-the-paris-agreement/

Greenpeace. (2015). Greenpeace response to final climate deal and EU contribution. Retrieved March 3, 2016, from http://www.greenpeace.org/eu-unit/en/News/2015/Paris-climate-deal-EU-performance/

Green, F., & Stern, N. (2016). China's changing economy: Implications for its carbon dioxide emissions. *Climate Policy*. doi:10.1080/14693062.2016.1156515

Gurría, A. (2015). *The climate challenge: Achieving zero emissions*. Lecture by the OECD Secretary-General. Retrieved March 3, 2016, from http://www.oecd.org/about/secretary-general/the-climate-challenge-achieving-zero-emissions.htm

Haites, E., Yamin, F., & Höhne, N. (2013). *Possible elements of a 2015 legal agreement on climate change*. Paris: IDDRI. Retrieved from http://www.iddri.org/Publications/Collections/Idees-pour-le-debat/WP1613_EH FY NH_legal agreement 2015.pdf

Höhne, N., Sterl, S., & Fekete, H. (2015). *How much more could Germany achieve through non-state action? Quantifying the impact of subnational and international cooperative initiatives on the future greenhouse gas missions of Germany*. Cologne: NewClimate Institute. Retrieved from https://newclimate.org/2015/11/25/how-much-more-could-germany-achieve-through-non-state-action/

Hulme, M. (2016). 1.5 °C and climate research after the Paris Agreement. *Nature Climate Change, 6*(3), 222–224. doi:10.1038/nclimate2939

Hulme, M., Wigley, T., Barrow, E., Raper, S., Centella, A., Smith, S., & Chipanski, A. (2000). *Using a climate scenario generator for vulnerability and adaptation assessments: MAGICC and SCENGEN version 2.4 workbook*. Norwich, UK: Climatic Research Unit, University of East Anglia. Retrieved from http://www.start.org/Projects/AIACC_Project/resources/ele_lib_docs/magicc_scengen_workbook.pdf

ICCT Europe. (2014). *European vehicle market statistics*. (Dr. Peter Mock, Ed.). Berlin: International Council on Clean Transportation. Retrieved from http://www.theicct.org/sites/default/files/publications/EU_pocketbook_2014.pdf

IEA. (2015). *Tracking clean energy progress 2015. Energy technology perspectives 2015 excerpt: IEA input to the clean energy ministerial*. Paris: International Energy Agency.

IPCC. (2013). *Climate change 2013: The physical science basis. Contribution of Working Group I to the Fifth Assessment Report of the Intergovernmental Panel on Climate Change*. (T. F. Stocker, D. Qin, G.-K. Plattner, M. Tignor, J. Boschung, A. Nauels, … P. M. Midgley, Eds.). Cambridge: Cambridge University Press.

IPCC. (2014). *Climate change 2014: Synthesis report. Contribution of Working Groups I, II and III to the Fifth Assessment Report of the Intergovernmental Panel on Climate Change*. (R. K. Pachauri & L. A. Meyer, Eds.). Geneva: IPCC.

IVM. (2015). *Non-state actors in a Paris agreement: are cities and companies bridging the ambition gap? A policy brief by IVEM and FORES*. Amsterdam: Institute for Environmental Studies, Free University Amsterdam. Retrieved from http://fores.se/wp-content/uploads/2015/05/NSA_Policy_brief_Bonn2.pdf

Jeffery, L., Fyson, C., Alexander, R., Gütschow, J., Rocha, M., Cantzler, J., … Blok, K. (2015). *2.7°C is not enough – we can get lower – Climate Action Tracker update 8 December 2015*. Berlin: Climate Action Tracker. Retrieved from http://climateactiontracker.org/assets/publications/briefing_papers/CAT_Temp_Update_COP21.pdf

Kallbekken, S., & Sælen, H. (2015). *Predicting Paris – Forecasting key outcomes from COP 21 using an expert survey*. Research note. Oslo, Norway: CICERO Center for International Climate and Environmental Research.

Kuramochi, T. (2015). Review of energy and climate policy developments in Japan before and after Fukushima. *Renewable and Sustainable Energy Reviews, 43*, 1320–1332. doi:10.1016/j.rser.2014.12.001

Kuramochi, T., Asuka, J., Fekete, H., Tamura, K., & Höhne, N. (2015). Comparative assessment of Japan's long-term carbon budget under different effort-sharing principles. *Climate Policy*. doi:10.1080/14693062.2015.1064344

Kuramochi, T., Wakiyama, T., & Kuriyama, A. (2015). *Comparative assessment of GHG mitigation scenarios for Japan in 2030* (IGES Working Paper WP1502). Hayama: Institute for Global Environmental Strategies (IGES) and Open Climate Network.

Lees, E. (2016). Responsibility and liability for climate loss and damage after Paris. *Climate Policy, 17*(1), 59–70. doi:10.1080/14693062.2016.1197095

Levin, K., & Fransen, T. (2015). *INSIDER: Why are INDC studies reaching different temperature estimates?* Washington DC, USA: World Resources Institute.

Mai, T., Logan, J., Blair, N., Sullivan, P., Bazilian, M., & Renewable, N. (2013). *RE-ASSUME: A decision maker's guide to evaluating energy scenarios, modeling, and assumptions*. A report for IEA Renewable Energy Technology Deployment programme. National Renewable Energy Laboratory.

Meinshausen, M., Raper, S. C. B., & Wigley, T. M. L. (2011). Emulating coupled atmosphere-ocean and carbon cycle models with a simpler model, MAGICC6–Part 1: Model description and calibration. *Atmospheric Chemistry and Physics, 11*(4), 1417–1456.

Meinshausen, M., Wigley, T. M. L., & Raper, S. C. B. (2011). Emulating atmosphere-ocean and carbon cycle models with a simpler model, MAGICC6–Part 2: Applications. *Atmospheric Chemistry and Physics, 11*(4), 1457–1471.

Metayer, M., Breyer, C., & Fell, H.-J. (2015). *The projections for the future and quality in the past of the World Energy Outlook for solar PV and other renewable energy technologies*. 31st European Photovoltaic Solar Energy Conference, September 14–18. doi:10.4229/31stEUPVSEC2015-7DV.4.61

Morgan, J., Dagnet, Y., Höhne, N., & Oberthür, S. (2014). *Race to the top: Driving ambition in the post-2020 international climate agreement* (ACT 2015 Working Paper). Washington DC: World Resources Institute. Retrieved from http://act2015.org/ACT2015_RacetotheTop.pdf

New Climate Economy. (2014). *Better growth, better climate*. The New Climate Economy Report. The Synthesis Report. Washington, DC: The Global Comission on the Economy and Climate. Retrieved from http://newclimateeconomy.report/TheNewClimateEconomyReport.pdf

NewClimate. (2015). *Second wave of climate change proposals (INDCs) expected in September after a first wave in March*. Retrieved February 28, 2016, from http://newclimate.org/2015/03/05/second-wave-of-climate-change-proposals-indcs-expected-in-september-after-a-first-wave-in-march

New Climate Economy. (2015). *Seizing the global opportunity: Partnerships for better growth and a better climate*. The 2015 New Climate Economy Report. Washington DC: The Global Commission on the Economy and Climate.

Oberthür, S., La Vina, A. G. M., & Morgan, J. (2015). *Getting specific on the 2015 climate change agreement: Suggestions for the legal text with an explanatory memorandum*. Retrieved from http://www.wri.org/sites/default/files/uploads/ACT2015_LegalSuggestions.pdf

Peters, G. P. (2016). The "best available science" to inform 1.5°C policy choices. *Nature Climate Change*. doi:10.1038/nclimate3000

Rogelj, J., & Knutti, R. (2016). Geosciences after Paris. *Nature Geoscience, 9*(3), 187–189. doi:10.1038/ngeo2668

Rogelj, J., Schaeffer, M., Friedlingstein, P., Gillett, N. P., van Vuuren, D. P., Riahi, K., ... Knutti, R. (2016). Differences between carbon budget estimates unravelled. *Nature Climate Change, 6*(3), 245–252. doi:10.1038/nclimate2868

Röser, F., Fekete, H., Höhne, N., & Kuramochi, T. (2015). *Options for assessing ambition of mitigation commitments beyond Paris*. Cologne. Retrieved from https://newclimateinstitute.files.wordpress.com/2015/12/background-paper-assessing-indcs-2015-12-01.pdf

Sharma, A. (2016). Precaution and post-caution in the Paris Agreement: Adptation, loss and damage, and finance. *Climate Policy, 17*(1), 33–47. doi:10.1080/14693062.2016.1213697

UK Government. (2008). *Climate change act 2008*. UK. Retrieved from http://www.legislation.gov.uk/ukpga/2008/27/contents

UNEP. (2015). *The emissions gap report 2015: A UNEP synthesis report*. United Nations Environment Programme (UNEP). Retrieved from https://newclimateinstitute.files.wordpress.com/2015/12/unep-emissions-gap-report-2015.pdf

UNFCCC. (1992). *United Nations Framework Convention on Climate Change* (ICOMM). Bonn: Author.

UNFCCC. (2015a). *Paris Agreement – Decision 1/CP.21 – Report of the Conference of the Parties on its twenty-first session, held in Paris from 30 November to 13 December 2015 Addendum Part two: Action taken by the Conference of the Parties at its twenty-first session*. Bonn. Retrieved from http://unfccc.int/resource/docs/2015/cop21/eng/10a01.pdf

UNFCCC. (2015b). *Synthesis report on the aggregate effect of the intended nationally determined contributions. FCCC/CP/2015/7*. Bonn: Author.

UNFCCC. (2016a). *Non-state actor zone for climate action platform*. United Nations Framework Convention on Climate Change. Retrieved from http://climateaction.unfccc.int/

UNFCCC. (2016b). *INDC portal*. Retrieved February 28, 2016, from http://unfccc.int/focus/indc_portal/items/8766.php

WRI. (2015): *STATEMENT: COP21 delivers "Historic" climate agreement in Paris*. Retrieved March 3, 2016, from http://www.wri.org/news/2015/12/statement-cop21-delivers-%E2%80%9Chistoric%E2%80%9D-climate-agreement-paris

Yamin, F., Haites, E., & Höhne, N. (2015). *From 90 Pages to 9: A possible Paris agreement from the Geneva negotiating text?* London. Retrieved from http://track0.org/wp-content/uploads/2015/06/Track-0-Legal-Agreement-9-pages-Monday-29-June-final.pdf

Precaution and post-caution in the Paris Agreement: adaptation, loss and damage and finance

ANJU SHARMA

Does the Paris Agreement constitute an adequate precautionary response to the current (un)certainties in keeping the global average temperature rise to safe levels? This article begins by briefly summarizing the scientific uncertainties and political challenges that the Paris Agreement is likely to face in meeting its goal. It then considers the adequacy of the Agreement's response to these uncertainties by assessing its key provisions related to dealing with climate impacts – adaptation, loss and damage and finance. It finds that these provision are weak in the face of very possible future realities. The interests of the poorest countries, communities and individuals and the most vulnerable, in particular, are inadequately served by the Paris Agreement.

Policy relevance

The article points out that the Paris Agreement does not satisfy the provisions of UNFCCC Article 3.3, which calls for a precautionary approach to addressing the adverse effects of climate change. Further action is needed to strengthen the provisions on mitigation and dealing with the impacts of climate change in the face of uncertainty.

Is the Paris Agreement (hereafter referred to as 'the Agreement') a breakthrough, as some commentators claim (Kinley, 2016; Roberts, 2015) or a failure, as others seem to think (Bawden, 2016)? The answer appears to depend on perspective. From the point of view of 'political realists', the Agreement pushes the boundaries of what was considered politically feasible, and is therefore a success. Politics is described simply as the art of the possible, however, and sometimes the possible is just not good enough. The real test of the Agreement lies in the extent to which it prepares the world, and in particular the poorest and most vulnerable, to deal with possible future realities.

The predecessor of the Agreement, the Kyoto Protocol, was considered the best possible outcome when it was agreed in 1997. Yet in the final analysis, although the countries that remained party to the Kyoto Protocol met their commitments (Shishlov, Morel, & Bellassen, 2016), it failed to make any appreciable dent in the upward trend of global GHG emissions (Schiermeier, 2012). Valuable time was lost as the Kyoto Protocol failed to wean the world off its fossil-fuel dependency, thus narrowing the possibility of limiting the impacts of global warming to a 'safe' level, and exposing countries and communities to greater risks.

An important lesson from the Kyoto Protocol experience should be to hope for the best (that the Agreement will succeed in meeting its goal of reducing GHG emissions and keeping global average temperature rise well below 2 °C compared with pre-industrial levels, and to pursue efforts to limit it to 1.5 °C), but prepare for the worst (take precautionary measures to minimize impacts and build resilience to deal with foreseen and unforeseen impacts even within the 2 °C envelope or for global temperatures overshooting safe levels).

The parent convention of the Paris Agreement, the UN Framework Convention on Climate Change (UNFCCC), calls for a precautionary approach in Article 3.3. It states that

> The Parties should take precautionary measures to anticipate, prevent or minimize the causes of climate change and mitigate its adverse effects. Where there are threats of serious or irreversible damage, lack of full scientific certainty should not be used as a reason for postponing such measures . . . (UNFCCC, 1992, p. 9)

Although Article 3.3 is mainly interpreted in the context of mitigation (addressing the causes of climate change), it also clearly calls for a precautionary approach to addressing the adverse effects of climate change (ILA, 2014; IPCC, 2001).

The role of a precautionary approach in adaptation to diminish and constrain possible harm is recognized in the academic literature, although the focus is mainly on the application of the approach to risk management strategies at the national level (Dessai & van de Sluijs, 2007) rather than on international policy-making. Heinzerling (2007), meanwhile, argues that the precautionary moment for action on climate change, when action was based on something less than a scientific consensus on the causes and consequences of climate change, has already passed. She argues that we are now in a 'post-cautionary' world, where climate change is already a reality. We must not only mitigate to avert the consequences of climate change, but also adapt to reduce harm from the consequences that we cannot avoid and deal with the losses and damages caused by the consequences that we cannot adapt to (Carlarne, Gray, & Tarasofsky, 2016).

There is already considerable scientific uncertainty regarding the extent to which the temperature goals agreed to in the Agreement can be considered 'safe'. In fact, it is almost certain that there will still be negative consequences – greater at 2 °C than at 1.5 °C. At the same time, there is considerable scientific and political uncertainty on whether the goals will be met. This gives rise to the question: does the Agreement take the precautionary and post-cautionary approach and account for this uncertainty in its provisions for dealing with climate change impacts? Do these provisions – mainly those related to adaptation, loss and damage and finance – constitute a strong enough response to cope with the possibility of a 1.5 °C, 2 °C or even greater rise in temperature?

Such a precautionary and post-cautionary approach is particularly important for the poorest and most vulnerable individuals, communities and countries, who are at the frontlines and are likely to be affected first and most by the impacts of climate change.

This article examines the Agreement's provisions on adaptation, loss and damage and finance to assess whether they are strong enough to deal with a 1.5 °C, 2 °C or even greater temperature rise. It is based on the premise that the robustness of these three key measures will indicate the extent to which the global community has taken a precautionary/post-cautionary approach on behalf of the world's most vulnerable. It begins by examining the temperature goals of the Agreement and summarizing the scientific uncertainties and political challenges that the Agreement is likely to face in meeting these goals.

1. How strong is the temperature 'defence line'?

In Article 2, the Agreement aims to hold the increase in the global average temperature to well below 2°C above pre-industrial levels and to pursue efforts to limit it to 1.5 °C. What is the extent of the impacts that we must prepare for, given these goals?

This '2 °C guardrail' was first proposed politically by the European Union in 1996 as a guide for mitigation efforts (New, Liverman, Schroder, & Anderson, 2010). It was thought to avoid serious risks when it was included in the Third Assessment Report of the Intergovernmental Panel on Climate Change (IPCC) in 2001, but has since been found to be inadequate to avoid serious risks to unique and threatened ecosystems and to avoid a large risk associated with extreme weather events (Richardson et al., 2009). It became a formal target in global negotiations only in 2010, under the Cancun Agreement. The choice of this number was not driven by scientific certainty that it will be a 'safe' goal, but more for reasons of economic effectiveness (Grubb, Hourcade, & Neuhoff, 2014) and through political consensus of what was considered to be achievable and tolerable (Knutti, Rogelj, Sedláček, & Fischer, 2016).

However, a 2 °C rise still poses considerable risks – particularly to small island states (through sea-level rise), and poor and vulnerable countries (through negative impacts on food and water security and livelihoods, for instance and through extreme weather events) (Schleussner et al., 2016). Recognizing this threat, the Alliance of Small Island States and the African Group proposed a goal of 'well below 1.5 °C' at the Copenhagen climate conference in 2009 (IISD, 2009). According to the IPCC, limiting the temperature rise to 1.5 °C will reduce the risk of dangerous sea level rise, but will still come with a high risk of extreme events and threats to ecosystems (IPCC, 2014) and significant impacts on poor communities. For instance, a 1.5 °C rise by 2030 could lead to 40% of present maize cropping areas in sub-Saharan Africa no longer being suitable for current cultivars, with strong repercussions on food security, economic growth and poverty reduction (World Bank, 2013).

Between June 2013 and February 2015, the UNFCCC organized four structured expert dialogues to consider the adequacy of the 2 °C goal in the light of the ultimate objective of the Convention (to prevent dangerous anthropogenic climate change); and the overall progress made towards achieving this goal. A technical summary report of these dialogues published in June 2015 recognized the high risks posed by a 2 °C rise (and, in some regions and vulnerable ecosystems, even for warming above 1.5 °C). Experts participating in the dialogues proposed restating the long-term global goal as a defence line or buffer zone, instead of a guardrail up to which all would be safe (UNFCCC, 2015a).

At the Paris conference, civil society groups argued that the inclusion of a 1.5 °C goal was a moral imperative, given the significant reduction of impacts compared with a 2 °C goal (Linnitt, 2015). Its final inclusion in the Agreement, even as an aspiration, came as a surprise to many, given strong opposition from countries such as India and Saudi Arabia. Although some attribute this success to the emergence of a 'high ambition coalition' at the Paris conference (Mathiesen & Harvey, 2015), others attribute it to a deal between some vulnerable states and the EU and US in exchange for language ruling out liability and compensation for loss and damage due to climate impacts in paragraph 51 of the decision adopting the Agreement (Raman, 2016).

Commentators warned against trading 'real benefits' that would be vital in a hotter and less stable future for an aspirational 1.5 °C goal (Goering, 2015), which they felt was already unachievable for scientific and political reasons.

1.1. Scientific challenges

Global average temperatures have already risen by 0.85 °C since 1880 (IPCC, 2014). According to the IPCC's Fifth Assessment Report, to hold warming below 2 °C, a 40–70% reduction in GHG emissions by 2050 (relative to 2010 levels) is required (IPCC, 2014).

To achieve the 1.5 °C aspiration, cuts of the order of 70–95% are needed by 2050 – requiring an immediate and radical transition instead of a mere fine tuning and the use of negative-emission technologies such as such as biomass energy with carbon capture and storage to suck back GHGs from the atmosphere. Even mitigation scenarios with a likely chance of keeping warming below 2 °C typically include an overshoot that will require the use of negative-emission technologies, which are currently neither available widely nor reliable (IPCC, 2014). These technologies come with their own risks – for instance, they could further endanger food security and impose a further stress on poor communities by endangering land rights (Anderson, 2015). Any slip-ups – a further delay of meaningful policy action or a failure to achieve negative emissions on a large scale – will make the 1.5 °C aim unattainable during this century (IIASA, 2015).

The emissions reductions currently offered by countries under the Agreement through their first set of Nationally Determined Contributions (NDCs), do not rise to the challenge of meeting the 1.5 °C aspiration or even the 2 °C goal. They will result in 55 gigatonnes (Gt) of GHG emissions in 2030, leading to a 3–4 °C rise by 2100 (UNEP, 2015). Paragraph 17 of the decision adopting the Agreement recognizes that the 55 Gt reduction through the NDCs will not meet the Agreement's temperature goal – emissions will have to fall to 40 Gt in 2030 to limit warming to 2 °C and lower to meet the 1.5 °C aim (UNFCCC, 2015b).

The Agreement includes a global stocktake every five years starting in 2023, and a facilitative dialogue in 2018, which could be used to drive up ambition in the NDCs – although this will not be easy. The Agreement's two-track approach, where some countries submit five-year NDCs and others ten-year NDCs, currently makes it difficult to aggregate contributions and assess collective ambition to compare it with where we should be to meet the Agreement's temperature goals (Müller, 2016a). Moreover, ambition will have to increased considerably even in the first NDCs – according to the Emissions Gap Report 2015 of the UN Environment Programme, all scenarios that lead to a 2 °C pathway require strong reductions after 2020 and rely on negative-emission technologies. Reduction rates will need to be even steeper for a 1.5 °C pathway (UNEP, 2015).

It is therefore not surprising, given these uncertainties, that the Agreement's 1.5 °C aspiration has been described by one scientist as 'whistling in the dark' (Hulme, 2016, p. 222). Others are even skeptical about whether the 2 °C aim can be met (Peters, 2015).

1.2. Political uncertainties

The extent to which a country can be legally bound under an international treaty is an interesting subject of debate. It is easier to convince countries to participate, become legally bound and to comply when they are responsive to levers that penalize or reward them – for instance, through unilateral trade sanctions (as in the case of the International Whaling Commission) or through positive incentives and assistance (as in the case of the Montreal Protocol).

However, it becomes trickier when participation or compliance is sought from more powerful countries that are immune to both carrots and sticks, as in the case of climate change. The US could not be forced or cajoled into participating in the Kyoto Protocol, despite the fact that it had sought major concessions during the negotiations that considerably weakened the overall effectiveness of the Kyoto Protocol. Although Canada was within legal boundaries to withdraw from the Kyoto Protocol, withdrawal proved to be virtually costless and without any serious repercussions. There was no question of sanctions, embargoes or 'retorsions' against Canada or the US, as there have been against Parties and non-Party States under the Convention on International Trade in Endangered Species of Wild Fauna and Flora (CITES), where most such sanctions – more than 95% – have been against developing countries (Sand, 2013).

Even within this widely acknowledged constraint in international law, however, the Agreement is remarkable for its leniency: countries have a legal obligation to submit NDCs under the Agreement and 'pursue domestic mitigation measures, with the aim of achieving the objectives of such contributions' (United Nations, 2015, Article 4.2, p. 4), but have no real obligation to fulfill them. This arrangement was a concession to the US, which wanted to keep out any language in the Agreement that could be considered legally binding to avoid having to submit it to a hostile Congress for ratification. As long as the Agreement was procedurally oriented, could be implementable on the basis of existing law and did not include legally binding emissions limits or new legally binding financial commitments, it could be ratified through a Presidential Executive Order instead (Bodansky, 2015).

As a result, any binding obligation to actually implement the NDCs was kept out of the Agreement. The extent to which this will achieve the intended result of keeping the US engaged in the global treaty and buying its participation is, however, far from certain (just as in the case of the Kyoto Protocol). When ratification is sought without ensuring broad political support, problems in implementation and future withdrawal, should the opposition come to power, are a distinct possibility. There are early signs of the difficulties that lie ahead in this regard even before the ink has dried on the Agreement: the US Supreme Court has stayed a critical piece of regulation that requires states to cut emissions from coal-fired power plants, which was central to the strategy of the US in getting the Agreement passed through a Presidential Executive Order (Liptak & Davenport, 2016).

Moreover, the compliance provisions of the Agreement are weaker than those of the Kyoto Protocol, with only a facilitative function. The Kyoto Protocol's compliance mechanism included an enforcement branch – although it is also worth noting that this was still ineffective in enforcing compliance by Canada, which instead chose to simply withdraw from the entire treaty.

2. The provisions to deal with climate impacts

Clearly, there is considerable uncertainty regarding the extent to which the temperature goals of the Agreement are safe and also the extent to which it is scientifically and politically possible to meet them. A precautionary and post-cautionary approach therefore calls for preparing the world to deal with impacts within and over the 2 °C goal.

To what extent does the Agreement prepare us to face this very real possibility? This section assesses the Agreement's provisions on adaptation, loss and damage and finance – all key elements in helping poor and vulnerable countries, communities and individuals in particular in dealing with climate impacts.

2.1. Adaptation

Adaptation is arguably both a precautionary and post-cautionary response to climate impacts – it can help to reduce the adverse effects of both scientifically certain (or already ongoing) and uncertain future climate impacts. Are the adaption-related provisions of the Agreement a strong enough global response for a 1.5 °C, 2 °C or even greater rise of global average temperatures?

Article 7 of the Agreement, on adaptation, establishes a global goal on adaptation:

> of enhancing adaptive capacity, strengthening resilience and reducing vulnerability to climate change, with a view to contributing to sustainable development and ensuring an adequate adaptation response in the context of the temperature goal referred to in Article 2. (United Nations, 2015, Article 7.1, p. 9)

This goal is not quantified and is more like a broad and general objective, which may not be very useful for establishing a clear and realistic roadmap for adaptation by itself. It does, however, link adaptation to the temperature goal, and hence to mitigation efforts. Article 7.4 further strengthens the link and includes a reference to adaptation costs: Parties recognize that 'greater levels of mitigation can reduce the need for additional adaptation efforts and that greater adaptation needs can involve greater adaptation costs' (United Nations, 2015, Article 7.4, p. 9).

Developing countries wanted to make this link to encourage mitigation ambition. However, it is debatable to what extent this will work in the absence of a clearly established responsibility for causing climate change and the application of the 'polluter pays' principle. The Agreement does not link a country's level of responsibility for causing climate change with a responsibility to pay for the impacts – therefore countries can continue to emit, confident that they cannot be held individually responsible or made to pay in proportion to their emissions.

In the context of Article 3.3 of the UNFCCC, which calls for a precautionary approach in addressing the adverse effects of climate change, it is also debatable whether the adaptation goal should have been linked to the 1.5 °C or 2 °C aims, when in fact current information indicates that temperatures could rise to 3–4 °C.

Developing countries wanted the Agreement to move beyond the focus on planning to the actual implementation of adaptation actions (ecbi, 2015). In the past, although some funds were made available for planning, funds for implementation were scarce – for instance, in the case of the National Adaptation Plans of Action (Huq, 2011). However, the focus remains once again on planning, assessing, sharing information and reporting. Countries are asked to engage in adaptation planning and to prepare adaptation communications where they can list priorities, implementation and support needs, as well as plans and actions.

To overcome the shortage of funds for adaptation, and to correct the imbalance between funds for mitigation and adaptation, developing countries had called for a quantitative goal for adaptation finance at the Paris meeting (UNFCCC, 2015c). No such goal was agreed and once again generalities relating to adaptation finance are repeated. Article 7.6 acknowledges the importance of 'support for and international cooperation on adaptation efforts... especially those that are most vulnerable' (United Nations, 2015, Article 7.6, p. 9) and Article 7.13 states: 'continuous and enhanced international support shall be provided to developing country Parties, to strengthen cooperation, engage in national planning processes and implementation of actions and to submit and update adaptation communications' (United Nations, 2015, Article 7.13, p. 11).

The existing provisions of the Agreement on adaptation, it would appear, do not constitute a strong enough precautionary or post-cautionary response to the uncertainties and certainties of climate impacts. However, as in the case of the mitigation ambition, the global stocktake process provides hope for enhancing adaptation action in the future: the adaptation communications provided by countries will form part of the stocktake, which is tasked with enhancing implementation of action, reviewing the adequacy of adaptation and support provided for adaptation and reviewing the overall progress made in achieving the global goal on adaptation (UNFCCC, 2015b). Adaptation is not specifically referenced in the context of the 2018 facilitative dialogue, however. The IPCC special report on the impacts of a 1.5 °C rise and related GHG emission pathways, due in 2018, could also provide a minimum baseline to judge the potential future flows of adaptation support and finance (Hulme, 2016).

2.2. Loss and damage

Loss and damage due to climate impacts could, arguably, be dealt with both through responses that are precautionary (for instance, insurance mechanisms) and post-cautionary (such as compensation for retrospective losses).

The very inclusion of a separate Article on loss and damage in the Agreement was a victory for developing countries in Paris – it was opposed until the very last minute by developed countries, who wanted loss and damage to be part of the Article on adaptation if at all it was to be included. They also wanted to explicitly exclude any association with liability and compensation (Adler, 2015). In the end, a stand-alone Article 8 on loss and damage was included, but paragraph 51 of the decision adopting the Agreement states that the Article 'does not involve or provide a basis for any liability or compensation' (UNFCCC, 2015b, p. 8, ¶ 51). As noted earlier, this was reported as being part of the closed-door deal between the US, EU and five small island developing states – the waiver was included in exchange for the separate Article on loss and damage and the inclusion of the 1.5 °C aim (Raman, 2016).

The lead negotiator for the Republic of Nicaragua, Paul Oquist, tried to protest its inclusion at the closing plenary (Doebbler, 2015). He is quoted as saying that 'this would mean giving a 3 °C temperature increase to our grandchildren and they are not able to ask for compensation. We strip them of any legal rights for legal action for the liability of other countries that caused the damage' (Raman & Chiew, 2015). It is not clear to what extent paragraph 51 will limit overall liability for climate change, although it is unlikely that a decision by the Conference of Parties (COP) to the UNFCCC can exclude the application of general rules on liability and compensation between states (Lees, 2016; Sharma et al., 2016).

Meanwhile, Article 8 ensures the continuation of the Warsaw International Mechanism for Loss and Damage and calls for cooperation and facilitation on areas such as, *inter alia,* early warning systems, emergency preparedness, insurance solutions and the resilience of communities, livelihoods and ecosystems. The decisions also request that the Warsaw International Mechanism Executive Committee establish a clearinghouse for risk transfer and a task force to develop recommendations to avert, minimize and address climate-related displacement (UNFCCC, 2015b).

There were no additional funds made available to deal with loss and damage, despite concerns expressed by developing countries that dipping into the already limited finance pool available for adaptation would shrink it further.

The loss and damage provisions of the Agreement would therefore appear to constitute very small first steps towards a precautionary and post-cautionary response to the certainties and uncertainties of current and future climate impacts. It is also not clear how or when these steps can be accelerated, as the Agreement does not provide for loss and damage to be part of the global stocktake.

2.3. Finance

Finance is a critical component of a global precautionary/post-cautionary response to deal with climate change impacts and a key indicator of the extent to which the global community is willing to act together to implement UNFCCC Article 3.3.

The finance outcome from the Paris meeting has been judged as particularly weak by some commentators (Müller, 2016b). To begin with, there is no clear quantitative commitment of finance in the Agreement itself. Instead, the decision adopting the Agreement extends the timeframe of the US$ 100 billion pledge, which was made in Copenhagen and formalized in Cancun, from 2020 to 2025. The COP serving as the Meeting of the Parties to the Paris Agreement (also called the CMA) will set a collective quantified goal for the post-2025 period from a floor of US$ 100 billion per year (UNFCCC, 2015b).

This US$ 100 billion figure has been controversial from the first time it was announced in Copenhagen. Developing countries have repeatedly sought clarity with regard to how the contributions are counted, the sources (public or private), additionality over development aid and the balance between adaptation and mitigation.

2.3.1. Counting climate finance

The controversy continued at the Paris meeting. Just before the conference, the Organisation for Economic Co-operation and Development (OECD), in collaboration with Climate Policy Initiative (CPI), published a report stating that public and private climate finance flows reached US$ 62 billion in 2014, with 77% addressing mitigation, 16% adaptation and 7% addressing both (OECD, 2015). These figures were contested in Paris – by both governments (Government of India, 2015) and civil society (Adaptation Watch, 2015). The latter found that 3,444 of the total 5,201 projects that were listed in the report as having either a principal or significant adaptation objective did not explicitly link the project activities to addressing climate vulnerability and US$7.7 billion of the US$10.1 billion claimed as adaptation finance in 2012 was not explicitly adaptation-related (Adaptation Watch, 2015).

To avoid such controversies in future, some developing country negotiators had proposed that a 'joint replenishment cycle' for the UNFCCC Financial Mechanism operated by the Global Environment Facility and the Green Climate Fund should be agreed on in Paris. While countries could continue to report private sector, development aid and bilateral contributions in their national communications and biennial assessments, the funds that flowed through the Financial Mechanism as part of this replenishment process would come with a surety against double-counting and provide a certain predictability (Müller & Ngwadla, 2015). This was not agreed.

There was, however, some progress made on enhancing the measurement, reporting and verification of climate finance. Developed countries are asked to report every two years on the finance they have provided and mobilized based on modalities, procedures and guidelines to be adopted by the Conference of the Parties serving as the meeting of the Parties to the Paris Agreement at its first session (CMA1). Other countries are encouraged to do so. They will also biennially communicate the projected levels of public finance that will be provided, if available.

Article 13 on transparency establishes a framework for transparency of support to provide clarity on the support provided and received for finance, mitigation, adaptation, capacity building and technology transfer – and to whatever extent is possible, provide a full overview of the aggregate financial support provided to inform the global stocktake. Developed countries and others that are providing support are asked to provide information on this support, whereas developing countries are asked to provide information on the support needed and received.

The global stocktake will also consider the information provided by developed country Parties and/or Agreement bodies on climate finance, to 'inform Parties in updating and enhancing, in a nationally determined manner, their actions and support' (United Nations, 2015, Article 14.3, p. 19). As noted earlier in the section on adaptation, it will also review the adequacy and effectiveness of adaptation and support provided – the Adaptation Committee, Least Developed Countries (LDC) Expert Group and Standing Committee on Finance have been tasked with making recommendations to CMA1 in this regard.

2.3.2. Additionality

The task of the Subsidiary Body for Scientific and Technological Advice (SBSTA), an ancillary body of the UNFCCC, which is to develop modalities for the accounting of financial resources, will be made all the more difficult by the omission of the phrase 'new and additional' from the Agreement. This standard phrase has been part of most climate agreements so far, including the 1997 Kyoto Protocol, the 2007 Bali Action Plan and the 2009 Copenhagen Accord. It is generally understood as being a reference to additionality to development aid commitments. Its omission from the Agreement could result in further disagreements in the future on what should or should not be counted as climate finance.

2.3.3. Sources

The question of whether climate finance is from public or private sector sources is particularly important in the context of dealing with the impacts of climate change, which tends to be non-profitable and hence cannot rely on private sources (Atteridge, 2011). The Agreement does not provide further clarity on whether climate finance pledges will be met through public or private sources. Article 9.3 states that 'developed country Parties should continue to take the lead in mobilizing climate finance from a wide variety of sources, instruments and channels', while merely 'noting the significant role of public funds' (United Nations, 2015, Article 9.3, p. 13). There is also no guarantee that grant-based funding will be provided to deal with climate impacts – the Agreement only states that the provision of finance should consider the need for grant-based resources for adaptation.

Developing countries also wanted a more trustworthy roadmap of how current levels of funding will be increased, instead of general assurances to scale up resources (UNFCCC, 2015c). Some of them proposed innovative or automatic sources of funding, such as the International Air Passenger Adaptation Levy proposed by the LDC (ecbi, 2008). Instead, Article 9.3 merely states that the mobilization of climate finance 'should represent a progression beyond previous efforts' (United Nations, 2015, Article 9.3, p. 13).

A surprise positive outcome from the Paris conference, however, is that a share of the proceeds from the new Sustainable Development Mechanism has been agreed, not only to cover the administrative expenses of the mechanism but also to meet the costs of adaptation in developing countries that are particularly vulnerable (UNFCCC, 2015b). This has been a divisive topic in the past, with developed countries unwilling to extend the share of proceeds that are applied to the Clean Development Mechanism under the Kyoto Protocol as an adaptation levy to Joint Implementation and emissions trading. The Agreement does not state what this share will be – this decision will be taken at CMA1.

2.3.4. Balance between adaptation and mitigation

There is considerable imbalance in the distribution of climate finance between mitigation and adaptation, as we saw in Section 2.3.1 – even according to the most optimistic figures (OECD, 2015), only 16% goes to adaptation. The need for a balance between mitigation and adaptation finance has been reiterated several times in climate agreements, including in the Copenhagen Accord and in the 2010 Cancun Agreement. This language itself has been described as a compromise – calling only for a balance with mitigation finance and not for adequate adaptation finance (Olhoff, Bee, & Puig, 2015). The costs of adaptation could be as high as US$280–500 billion per annum by 2050 according to a recent report– considerably more than 50% of US$100 billion (UNEP, 2016).

The Agreement simply reiterates the need for balance. Article 9.4 states that the 'provision of scaled-up financial resources should aim to achieve a balance between adaptation and mitigation' (United Nations, 2015, Article 9.4, p. 13). However, this language is likely to stay as ineffective as in the past in the absence of a quantitative target for adaptation alone or clarity about what this balance means, how it should apply to the various sources of climate finance and how it can be ensured.

3. Still far to go

At the Paris conference vulnerable countries focused on getting the 1.5 °C goal into the Agreement because a 2 °C rise posed significantly greater danger to them. They found support from the EU and the US. A skeptic could argue that it was easier for the countries responsible for climate change to give in to an aspirational goal than to heed the precautionary principle and acknowledge that it was time to prepare for a much more dangerous rise in global average temperatures. This would have required a much more robust and precautionary response, with stronger provisions to help countries adapt, deal with loss and damage and manage costs.

Irrespective of whether a trade was made between the inclusion of the lower aspirational temperature goal and the sentence excluding liability and compensation in the Paris decision, the Agreement leaves vulnerable countries, and poor and vulnerable communities and individuals around the world, exposed and with very little by way of tangible support. Several concrete proposals that would have constituted a more robust response to climate impacts, and were in the draft negotiating text, were dropped at the last minute in Paris. These include proposals to: plan and undertake adaptation on the basis of an evaluation of temperature scenarios resulting from aggregate mitigation action; ensure that the level of support meets the needs for adaptation; set quantitative targets for adaptation finance; ensure that climate finance is predictable, new and additional; provide additional support for loss and damage; and to set up a compensation regime (UNFCCC, 2015c).

As the costs of dealing with climate impacts rise, a more concrete and fair method of sharing these costs between countries on the basis of responsibility is becoming more and more urgent. Solutions that are not considered fair or just by many countries are unlikely to be sustainable or effective in the long term. The adaptation goal makes a very small start towards creating a link between mitigation action and adaptation, but it is ineffective without a concrete link to a country's responsibility to pay for impacts. There have been stronger proposals on the table.

For instance, Brazil proposed a Clean Development Fund financed through penalties on countries who fail to comply with their legally binding emission targets, decided according to the impact of their cumulative historical emissions starting from 1840, on the global-average surface temperature (UNFCCC, 1996). This proposal was put forward before the Kyoto conference in 1997, but was referred to SBSTA, which continued to discuss it under an *ad hoc* group for the modelling and assessment of historic contributions to climate change (MATCH). MATCH concluded that scientific underpinnings, historical datasets and modelling tools are now available to evaluate contributions to climate change with some scientific confidence by regional, national or sectoral breakdown, except for emissions from land-use change and forestry (Höhne et al., 2008). In 2006, SBSTA concluded that MATCH had established robust methodologies to quantify historical contributions to climate change.

After Kyoto and in the run up to the Copenhagen and Paris meetings, there have been other studies that have considered differentiated responsibilities for emissions and also differentiated capabilities that could form the basis of assigning responsibility and share the costs of impacts (Müller, Höhne, & Ellermann, 2007; Müller & Mahadeva, 2014). Others have proposed creating a voluntary Compensation Fund for Climate Impacts with specialized, independent climate courts to adjudicate cases and determine compensation payments based on precedents in other multilateral negotiations such as the International Oil Pollution Compensation Funds or the International Nuclear Liability regime (Sprinz & Von Bünau, 2013).

Until now, however, the countries that have contributed the most to climate change are far more interested in sharing the costs of mitigation than they are in sharing the costs of adaptation and of dealing with climate change impacts (Shue, 2014).

4. Conclusion

Nearly two decades ago, when the Kyoto Protocol was adopted, it was greeted with as much enthusiasm as the Agreement. Those who felt that the Kyoto Protocol did not go far enough to disincentivize fossil fuel use and trigger a move to renewables were told that it was a first step, the best that was possible within the political realities of the day and that efforts would be strengthened over time. Adaptation was left on the back-burner as the world staked its hopes on mitigation.

Twenty years later the urgency of intensifying our responses to climate impacts is stark, but it would seem that we have not learnt the value of precaution. The interests of the poorest and most vulnerable, in particular, are inadequately served by the Agreement.

The Agreement has been praised for giving adaptation as much attention as mitigation. On closer scrutiny, however, and in light of the precautionary principle, the adaptation provisions fall well short. Loss and damage has not received the importance it should have at this stage, when countries are already sustaining losses due to climate impacts. The Agreement's provisions on climate finance, an easily measureable indicator of global cooperation, do not currently inspire confidence. But time is running out. Climate change threatens to push more than 100 million people back into poverty over the next fifteen years (World Bank, 2015). We cannot afford the luxury of going back to the drawing board for another round of fraught negotiations.

There are actions within and outside the Agreement that we can build on to fortify the weaknesses of the Agreement. For instance, the goal of 1.5 °C may become more realistic and achievable if the 2018 facilitative dialogue becomes a genuine opportunity to compare aggregate emissions reductions included in the NDCs with where the IPCC Special Report tells us we should be to achieve the more ambitious temperature aspiration and to assess whether countries will actually ramp up ambition in their NDCs to close the gap.

The discussion on adaptation has to move on from the focus on planning to full-scale implementation, with adequate financial and other means of implementation provided at the global level. The global willingness to share responsibility for adaptation must go beyond words and plans – mutually agreeable ways of assessing adaptation needs and raising adequate support have to be found.

The discussion on transparency related to climate finance is important, but it is also very erosive of trust. Ways of building mutual trust in this regard have to be found – for instance, through the proposed joint replenishment cycle for the UNFCCC Financial Mechanism mentioned earlier, which to some extent overcomes the problems of defining what may or may not count as climate finance and also ensures a certain amount of predictability. There is already a precedent under the Montreal Protocol, where replenishments to the Multilateral Fund are decided on the basis of a needs assessment carried out by the Protocol's Technology and Economic Assessment Panel (Müller & Ngwadla, 2015).

Finally, the discussion on dealing with the costs of loss and damage due to climate change cannot be wished away forever, especially if global average temperatures overshoot the goal in the Agreement. A strong and early precautionary recognition of this could help to drive a more committed response to both mitigation and adaptation, perhaps serving as the carrot and stick that is currently so elusive in the climate negotiations.

Disclosure statement

No potential conflict of interest was reported by the authors.

References

Adaptation Watch. (2015). *The 2015 adaptation finance transparency gap report.* Retrieved from http://static1.squarespace.com/static/56410412e4b09d10c39ce64f/t/56552e96e4b0f60cdb91b6ac/1448423062880/AW_Report_24+11+15.pdf

Adler, B. (2015, December 8). Here's why the words "loss and damage" are causing such a fuss at the Paris climate talks. *Grist.* Retrieved from http://grist.org/climate-energy/heres-why-the-words-loss-and-damage-are-causing-such-a-fuss-at-the-paris-climate-talks/?utm_source=syndication&utm_medium=rss&utm_campaign=feed

Anderson, T. (2015). *Caught in the net: How "net-zero emissions" will delay real climate action and drive land grabs*. Retrieved from http://www.actionaid.org/sites/files/actionaid/caught_in_the_net_actionaid.pdf

Atteridge, A. (2011). *Will private finance support climate change adaptation in developing countries? Historical patterns as a window on future private-sector climate finance* (Working Paper No. 2011–05). Retrieved from http://www.sei-international.org/mediamanager/documents/Publications/Climate/SEI-WP-2011-05-Private-Sector-Adaptation-Finance-ES.pdf

Bawden, T. (2016, January 8). COP21: Paris deal far too weak to prevent devastating climate change, academics warn. *Independent*. Retrieved from http://www.independent.co.uk/environment/climate-change/cop21-paris-deal-far-too-weak-to-prevent-devastating-climate-change-academics-warn-a6803096.html. Retrieved from http://www.c2es.org/docUploads/legal-options-us-acceptance-new-climate-change-agreement.pdf

Bodansky, D. (2015). *Legal options for U.S. acceptance of a new climate change agreement*. Retrieved from http://www.c2es.org/docUploads/legal-options-us-acceptance-new-climate-change-agreement.pdf

Carlarne, C. P., Gray, K. R., & Tarasofsky, R. G. (Eds.). (2016). *The Oxford handbook of international climate change law*. Oxford: Oxford University Press.

Dessai, S., & van de Sluijs, J. (2007). *Uncertainty and climate change adaptation: A scoping study*. Retrieved from http://s3.amazonaws.com/academia.edu.documents/30786964/NWS-E-2007-198.pdf?AWSAccessKeyId=AKIAJ56TQJRTWSMTNPEA&Expires=1467277927&Signature=pnoqDvLNzqrCstJd%2B78tRrhxcns%3D&response-content-disposition=inline%3B%20filename%3DUncertainty_and_climate_change_adaptatio.pdf

Doebbler, C. F. J. (2015, December 15). Paris climate deal: How could they do this to us? *Counterpunch*. Retrieved from http://www.counterpunch.org/2015/12/15/paris-climate-deal-how-could-they-do-this-to-us

ecbi. (2008). *International air passenger adaptation levy (IAPAL): A proposal by the group of least developed countries (LDCs) within the framework of the Bali action plan*. Retrieved from http://www.eurocapacity.org/downloads/ecbi_Brief_-_IAPAL_13_Q&As.pdf

ecbi. (2015). *Oxford seminar 2015*. Retrieved from http://www.eurocapacity.org/downloads/2015_Oxford_Seminar_report.pdf

Goering, L. (2015, December 11). Poor states warned not to trade away protection at climate talks. *Reuters*. Retrieved from http://www.reuters.com/article/climatechange-summit-goal-idUSL8N1403DZ20151211

Government of India. (2015). *Climate change finance, analysis of a recent OECD report: Some credible facts needed*. Retrieved from http://pibphoto.nic.in/documents/rlink/2015/nov/p201511291.pdf

Grubb, M., Hourcade, J.-C., & Neuhoff, K. (2014). Trapped? In *Planetary economics: Energy, climate change and the three domains of sustainable development* (pp. 1–45). Abingdon: Routledge.

Heinzerling, L. (2007). *Climate change, human health, and the post-cautionary principle* (O'Neill Institute Paper 4). Retrieved from http://scholarship.law.georgetown.edu/ois_papers/4

Höhne, N., Penner, J., Prather, M., Fuglestvedt, J., Lowe, J., & Hu, G. (2008). *Summary report of the ad hoc group for the modelling and assessment of contributions to climate change (MATCH)*. Retrieved from http://unfccc.int/files/methods/other_methodological_issues/application/pdf/match_summary_report_.pdf

Hulme, M. (2016). 1.5 °C and climate research after the Paris Agreement. *Nature Climate Change, 6*, 222–224.

Huq, S. (2011). *Adaptation: Resources now to plan and implement*. Retrieved from http://pubs.iied.org/pdfs/17117IIED.pdf

IIASA. (2015). *What would it take to limit climate change to 1.5°C?* Retrieved from http://www.iiasa.ac.at/web/home/about/news/150521-15-scenarios.html

IISD. (2009). Summary of the Copenhagen climate change conference: 7–19 December 2009. *Earth Negotiations Bulletin, 12*, 459.

ILA. (2014). *Declaration of legal principles relating to climate change: Resolution 2/2014*. Retrieved from http://www.ila-hq.org/en/committees/index.cfm/cid/1029

IPCC. (2001). Section 10.4.2.2, Precautionary Considerations. In *Climate change 2001: Mitigation* (p. 655). Contribution of Working Group III to the Third Assessment Report of the Intergovernmental Panel on Climate. Retrieved from https://www.ipcc.ch/ipccreports/tar/wg3/pdf/10.pdf

IPCC. (2014). *Climate change 2014: Synthesis report.* Contribution of Working Groups I, II and III to the Fifth Assessment Report of the Intergovernmental Panel on Climate Change. Retrieved from http://www.ipcc.ch/pdf/assessment-report/ar5/syr/SYR_AR5_FINAL_full_wcover.pdf

Kinley, R. (2016). Climate change after Paris: From turning point to transformation. *Climate Policy, 17*(1), 9–15. doi:10.1080/14693062.2016.1191009.

Knutti, R., Rogelj, J., Sedláček, J., & Fischer, E. M. (2016). A scientific critique of the two-degree climate change target. *Nature Geoscience, 9*, 13–18.

Lees, E. (2016). Responsibility and liability for climate loss and damage after Paris. *Climate Policy, 17*(1), 59–70. doi:10.1080/14693062.2016.1197095.

Linnitt, C. (2015). Global leaders fight for new 1.5 degrees warming target at COP21 climate talks. Retrieved from http://www.desmogblog.com/2015/12/01/global-leaders-fight-new-1-5-degrees-warming-target-cop21-climate-talks

Liptak, A., & Davenport, C. (2016, February 9). Supreme court deals blow to Obama's efforts to regulate coal emissions. *The New York Times*. Retrieved from http://www.nytimes.com/2016/02/10/us/politics/supreme-court-blocks-obama-epa-coal-emissions-regulations.html?_r=0

Mathiesen, K., & Harvey, F. (2015, December 8). Climate coalition breaks cover in Paris to push for binding and ambitious deal. *The Guardian*. Retrieved from http://www.theguardian.com/environment/2015/dec/08/coalition-paris-push-for-binding-ambitious-climate-change-deal

Müller, B. (2016a). *From contribution framework to ambition mechanism: How to enhance mitigation ambition under the Paris Agreement*. Retrieved from http://www.eurocapacity.org/downloads/Dynamic_Ambition_Mechanism_published.pdf

Müller, B. (2016b). Finance in Paris: Non à la nouvelle haute couture impériale! *Oxford Climate Policy Blog*. Retrieved from http://oxfordclimatepolicy.com/blog/finance-in-paris

Müller, B., Höhne, N., & Ellermann, C. (2007). Differentiating (historic) responsibilities for climate change. *Climate Policy, 9*, 593–611.

Müller, B., & Mahadeva, L. (2014). *The Oxford approach: Operationalizing the UNFCCC principle of 'respective capabilities'*. Retrieved from http://www.oxfordclimatepolicy.org/publications/documents/TheOxfordApproach_ecbiBrief.pdf

Müller, B., & Ngwadla, X. J. (2015). *A Paris replenishment cycle for contributions to the UNFCCC financial mechanism*. Retrieved from http://www.oxfordclimatepolicy.org/publications/documents/Replenishing_the_FM__final.pdf

New, M., Liverman, D., Schroder, H., & Anderson, K. (2010). Four degrees and beyond: The potential for a global temperature increase of four degrees and its implications. *Philosophical Transactions of the Royal Society A, 369*, 6–19.

OECD. (2015). *Climate finance in 2013–14 and the USD 100 billion goal*. Retrieved from http://climatepolicyinitiative.org/wp-content/uploads/2015/10/OECD-2015-Climate-Finance-Report-with-CPI-LR.pdf

Olhoff, A., Bee, S., & Puig, D. (2015). *The adaptation gap report update with insight from the INDCs*. Retrieved from http://web.unep.org/sites/default/files/gapreport/UNEP_Adaptation_Finance_Gap_Update.pdf

Peters, G. (2015). *A journey from 5 °C to 2 °C*. Retrieved from http://www.cicero.uio.no/en/posts/news/a-journey-from-5c-to-2c

Raman, M. (2016). *The climate change battle in Paris: An initial analysis of the Paris COP21 and the Paris agreement*. Retrieved from http://www.twn.my/title2/climate/doc/epw%20arrticle%20ref%202.pdf

Raman, M., & Chiew, H. (2015, December 21). Paris Agreement adopted with "corrections". *Social watch*. Retrieved from http://www.socialwatch.org/node/17118

Richardson, K., Steffen, W., Schellnhuber, H. J., Alcamo, J., Barker, T., Kammen, D. M., … Wæver, O. (2009). *Climate change—global risks, challenges & decisions: Synthesis Report*. Denmark: Copenhagen University. Retrieved from https://www.pik-potsdam.de/news/press-releases/files/synthesis-report-web.pdf.

Roberts, D. (2015, December 15). The conceptual breakthrough behind the Paris climate treaty. *Vox*. Retrieved from http://www.vox.com/2015/12/15/10172238/paris-climate-treaty-conceptual-breakthrough

Sand, P. (2013). Enforcing CITES: The rise and fall of trade sanctions. *Review of European Community & International Environmental Law, 22*, 251–263.

Schiermeier, Q. (2012). The Kyoto Protocol: Hot air. *Nature*, *491*, 656–658.

Schleussner, C., Lisssner, T. K., Fischer, E. M., Wohland, J., Perrette, M., Golly, A., . . . Schaeffer, M. (2016). Differential climate impacts for policy-relevant limits to global warming: The case of 1.5°C and 2°C. *Earth System Dynamics*, *7*, 327–351.

Sharma, A., Schwarte, C., Müller, B. Abeysinghe, A., & Barakat, S. (2016). *Pocket guide to the Paris Agreement*. Oxford: European Capacity Building Initiative.

Shishlov, I., Morel, R. and Bellassen, V. (2016). Compliance of the parties to the Kyoto protocol in the first commitment period. *Climate Policy*, *16*(6), 768–782. doi:10.1080/14693062.2016.1164658.

Shue, H. (2014). *Climate justice: Vulnerability and protection*. Oxford: Oxford University Press.

Sprinz, D. D., & Von Bünau, S. (2013). The compensation fund for climate impacts. *Weather, Climate and Society*, *5*, 210–220.

UNEP. (2015). *Executive summary of the 2015 emissions gap report*. Retrieved from http://uneplive.unep.org/media/docs/theme/13/EGR_2015_ES_English_Embargoed.pdf

UNEP. (2016). *The adaptation gap finance report*. Retrieved from http://drustage.unep.org/adaptationgapreport/sites/unep.org.adaptationgapreport/files/documents/agr2016.pdf

UNFCCC. (1992). *United Nations framework convention on climate change*. Retrieved from http://unfccc.int/resource/docs/convkp/conveng.pdf

UNFCCC. (1996). *Implementation of the Berlin Mandate: Additional proposals from Parties. FCCC/AGBM/1997/MISC.1/Add.3*. Retrieved from http://unfccc.int/cop5/resource/docs/1997/agbm/misc01a3.htm

UNFCCC. (2015a). *Report on the structured expert dialogue on the 2013–2015 review*. Retrieved from http://unfccc.int/resource/docs/2015/sb/eng/inf01.pdf

UNFCCC. (2015b). *Report of the Conference of the Parties on its twenty-first session, held in Paris from 30 November to 13 December 2015. Addendum. Part Two: Action taken by the Conference of the Parties at its twenty-first session* (FCCC/CP/2015/10/Add.1). Paragraph 17. Retrieved from http://unfccc.int/resource/docs/2015/cop21/eng/10a01.pdf

UNFCCC. (2015c). *Ad hoc working group on the Durban platform for enhanced action: Second session, part eight* (FCCC/ADP/2015/1). Retrieved from http://unfccc.int/resource/docs/2015/adp2/eng/01.pdf

United Nations. (2015). *Paris Agreement*. Retrieved from http://unfccc.int/files/essential_background/convention/application/pdf/english_paris_agreement.pdf

World Bank. (2013). *Turn down the heat: Climate extremes, regional impacts, and the case for resilience*. Retrieved from http://documents.worldbank.org/curated/en/2013/06/17862361/turn-down-heat-climate-extremes-regional-impacts-case-resilience-full-report

World Bank. (2015). *Shock Waves: Managing the impacts of climate change on poverty*. Retrieved from https://openknowledge.worldbank.org/bitstream/handle/10986/22787/9781464806735.pdf?sequence=13&isAllowed=y

The Paris Agreement: China's 'New Normal' role in international climate negotiations

ISABEL HILTON, AND OLIVER KERR

The political success of the Paris COP 21 in 2015 contrasted sharply with the diplomatic breakdown at Copenhagen six years before in 2009. Contributing to this success was China's emergence as a more positive participant in the international climate change negotiations. Despite efforts to reduce energy and carbon intensity since the mid-2000s, Chinese negotiators in Copenhagen were careful not to link domestic action on climate change to any presumptions of international obligation. This stands in marked contrast to Paris where, for the first time, China was willing to commit to an absolute cap on emissions subject to international measurement, reporting, and verification (MRV). This article examines what changed between 2009 and 2015 such that China was able to play a more constructive role in the global climate regime. It argues that a key driver of change was China's shift to a 'New Normal' model of economic development. Beginning in the 12th Five Year Plan Period (2011–2015), China's economic policy prioritized a transition from energy-intensive growth based on heavy industry, exports and investment, to a more balanced economy characterized by slower growth, an increasing role for services and domestic consumption, and a focus on innovation and low-carbon technologies. This transition gave China the opportunity to re-formulate its priorities in international climate negotiations and helped pave the way for increased climate cooperation with the US, the lack of which had been a major roadblock to success in Copenhagen. Progress was further facilitated by a range of external factors, including impressive French diplomacy in the run-up to COP21 and the important shift to a bottom-up, voluntary approach to commitments.

Policy relevance

This article analyses the reasons behind China's more constructive engagement at COP 21 in Paris, which contributed to the Conference's successful outcome. It is important for policy makers to better comprehend China's shifting position and the motivations behind it, given the country's key role in keeping temperature rise at a safe level. The deal reached in Paris now needs to be implemented, and success is by no means guaranteed. A deeper understanding of the position of this key player can only help with the negotiation process, implementation phase, and further negotiations that are needed to strengthen targets.

At the 21st Conference of the Parties, in Paris in December 2015, 195 countries adopted a comprehensive agreement to reduce carbon emissions to net zero by the second half of the century. Whether this ambition will be matched with action is a matter of some debate (Clémençon, 2016), but arguably the Paris Agreement's success lies in setting out a framework on which plans can be built, and in the transformation of the climate narrative from one of economic burden and political failure to a story of opportunity (Christoff, 2016).

This is in marked contrast to the outcome of COP15 in Copenhagen in 2009. Failure then to reach agreement cast a long shadow over subsequent attempts to move climate diplomacy forward. A common theme addressed in many post-mortem examinations of Copenhagen is the role played by China (Christoff, 2010; Conrad, 2012). While some (Bodansky, 2010) have been more positive about China's contribution, reports in the US and European media tended to note China's perceived diplomatic obstructionism (Revkin & Broder, 2009) and accused China of sabotaging the negotiations and refusing to back higher ambition, even where it had no direct impact on its own obligations (Lynas, 2009).

In the run-up to the Paris conference, one of the key questions was whether China would play a more constructive role than in 2009. It was all the more striking, then, that China emerged from the Paris COP with a reputation for climate cooperation that had survived the negotiations relatively unscathed (AP, 2015; Li, 2016; Ye & Wu, 2015). What had changed, between Copenhagen and Paris, to bring this about?

This article argues that a key driver of change has been China's shift to a 'New Normal' model of economic development, beginning in the 12th Five Year Plan Period (2011–2015), which emphasizes a move away from energy-intensive industry, exports and investment-led growth, and toward services and domestic consumption, with a focus on low-carbon technologies. This gave China the opportunity to re-formulate its priorities in international climate negotiations and helped pave the way for increased climate cooperation with the US, the lack of which had been a major roadblock to success in Copenhagen. Progress was further facilitated by a range of other factors, including impressive French diplomacy in the run-up to COP21 and the important shift to a bottom-up, voluntary approach to commitments.

1. Domestic action, global inaction

China's domestic commitment to tackling climate change and associated environmental challenges pre-dates Copenhagen (Li, 2016). As early as 2004, China's Medium and Long Term Energy Conservation Plan for 2020 outlined a range of measures to promote energy efficiency, including the country's first energy intensity reduction targets (NDRC, 2004). This was followed in 2005 by a renewable energy law, while binding national targets on energy intensity and the reduction of major pollutants were introduced in the 11th Five Year Plan (2006–2010). Then, in 2007, China set up its national lead office on climate change and issued the first national climate change plan of any developing country (Geall & Hui, 2015). By the time the Chinese negotiators came to Copenhagen, they were able to point to a pledge to reduce China's emissions intensity by 40–45% by 2020. Although it was not a commitment to peak or reduce total emissions, it was one of the more ambitious proposals from an emerging economy.

At the same time, policy makers were careful not to link domestic action on climate change to any presumptions of international obligation, and at Copenhagen Chinese negotiators rejected any binding emissions target that, they argued, could potentially serve as a constraint on development. Conrad (2012) notes the apparent 'contradiction' between China's domestic ambition with respect to climate policy and its more ambivalent negotiating position on the international stage in 2009. This stands in marked contrast to 2015 in Paris, where China's domestic progress on tackling climate change was reflected in a willingness to commit, for the first time, to emissions reductions that would be subject to international measurement, reporting, and verification (MRV).

The link between domestic politics and international relations has long been recognized (Putnam, 1988). The shift in China's diplomatic position in international climate negotiations cannot, however, be explained only with reference to China's domestic environmental policy agenda. Rather, we must also look to changes in China's economic policy to explain the shift in approach between Copenhagen and Paris.

2. Copenhagen climate negotiations in an era of industry-led growth, 2000–2013

From 2000 to 2013, Chinese economic policy was characterized by high levels of investment and exports, with a focus on energy-intensive heavy industry.[1] GDP growth averaged 10.5% each year, with manufacturing of commodities like cement and steel contributing to high levels of primary energy and coal consumption, both of which grew at an average rate of around 8% per year – as did carbon dioxide emissions (Green & Stern, 2016; UNFCCC and Boden, Marland, & Andres, 2015). Consequently, even though this period saw China lay the foundations of its climate policy, take steps to promote industrial efficiency, and reduce the energy intensity of GDP, there was reluctance to subscribe to any international climate regime that threatened to undermine its energy-intensive growth model. In short, climate change took second place to the more urgent domestic imperative of continuing to deliver the prosperity that was the foundation of the party-state's credibility after 1989.

This perceived trade-off between economic development and climate mitigation was embedded in the structure of the negotiations themselves. In particular, the division between Annex I and non-Annex I countries, set up in the 1992 United Nations Framework Convention on Climate Change (UNFCCC) itself, could not easily accommodate emerging economies like China that increasingly fell neatly into neither category. In 2009, China still positioned itself as a primarily innocent victim of climate change, a poor country, albeit a large one, that spoke for and defended the interests of other developing (non-Annex I) countries.

With only modest GDP on a per capita basis, China found in its non-Annex I status a ready-made justification for its unwillingness to sign up to binding international commitments. At the same time, by 2006, China's emissions had already surpassed those of the US, and, due largely to a substantial stimulus package launched in response to the 2008 financial crisis, China's GDP at the end of 2010 had returned to double digits at 10.6% (National Bureau of Statistics (NBS), 2016b). In the US, domestic climate scepticism and political opposition to robust climate mitigation was further fuelled by the belief that China was getting away with growing its emissions, while shifting an unfair burden on the US. In China, on the other hand, US insistence that China must also reduce its emissions was seen as a US effort to slow or impede China's legitimate development. This fundamental difference of views led to a deadlock in Copenhagen, broken in Paris after transformational changes in China's economic policy landscape during the 12th Five Year Plan (2011–2015) (Grubb et al., 2015).

3. The road to Paris: restructuring and the 'New Normal' economy, 2014–2015

Since Copenhagen and particularly beginning in 2011, there has been a steady increase in both the scope and ambition of China's domestic climate policy. The 12th Five Year Plan included clear energy and carbon intensity targets, investment in low-carbon industries, increased targets for R&D spending, pilot carbon markets, and measures to boost renewables in order to reduce dependency on coal in the energy mix (Zhang, 2015). These measures, however, must be seen in the context of a radical shift in economic policy, which in the 12th FYP period became linked explicitly with the climate agenda (Stern, 2011; State Council of the People's Republic of China, 2014).

There has long been recognition in China that double-digit growth was unsustainable, in both environmental and economic terms. Come the 12th FYP in 2011, the balance of China's policy had begun to change: the costs to China's population of the model of development were tangible in worsening air, water, and soil pollution, even as the rewards of the model were shrinking. China was beginning to feel the pressure of diminishing returns as rising labour costs drove up production costs, and over-investment began to generate surplus capacity and redundant developments rather than growth (Cai, 2012; Garnaut & Huang, 2005; Zhu & Kotz, 2011). China's planners could see that systemic change was inevitable (Li & Wang, 2012), with a move up the value chain and less waste in energy and resources.

The Third Plenary Session of the 18th Chinese Communist Party Congress in November 2013 signalled a marked change in economic policy compared to the previous decade (CCCPC, 2013). The new model of development, fleshed out by new Chinese leadership under President Xi Jinping and Premier Li Keqiang over the following two years, was to be characterized by slower growth, a gradual transition from heavy industry to services, increased productivity through innovation, and a greater role for domestic consumption to decrease over-reliance on investment (NBS, 2015b; PRC, 2011; Yan, 2015). Targets were introduced for slower economic growth at 7% per year and for services to account for 47% of value added by 2015, up from 43% in 2010 (PRC, 2011), to be achieved through a suite of policy measures including investment, financial sector reform, taxes, and the removal of subsidies for energy-intensive industries and exports (Qi, Winchester, Karplus, & Zhang, 2016).

Environmental sustainability was also prioritized, no longer seen as an issue separate from the economic policy but inextricably linked to 'accelerating economic restructuring' (State Council of the People's Republic of China, 2014). Emissions reduction would be both a natural outcome of slower growth and a more balanced economy, in addition to something to be explicitly promoted through targeted policy (Green & Stern, 2016). By 2015, Xi Jinping talked of a 'New Normal' in China, with 'ecological civilization' elevated to the status of a key slogan that would inform future planning (Xinhua, 2015).

4. A 'New Normal' for international climate cooperation

In the run-up to Paris, the effects of China's new development model were already visible. According to official figures, GDP growth declined to 7.3% in 2014 and 6.9% in 2015 (Figure 1), a marked change from the double-digit growth that had defined the previous decade. The was also evidence of a gradual shift away from industry and manufacturing (Figure 2), while in 2015 the contribution of

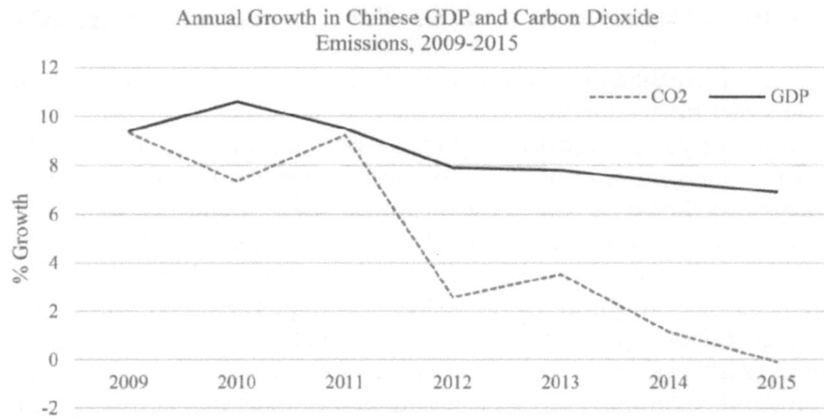

Figure 1 Slowing GDP and emissions growth, 2009–2016
Sources: GDP data from NBS (2016b) Online Database. Emissions data calculated from UNFCCC (2015). 2015 growth rate taken from Korsbakken et al. (2016a).

services to total value added rose to over 50% for the first time, exceeding government targets (PRC, 2016).[2] That same year, steel output dropped by 2.2%, cement production by 5.3%, and new construction by 14% in terms of floor space (NBS, 2016a). Together with continued efforts to promote efficiency and diversify the energy mix, these trends had two main impacts. First, growth in total primary energy consumption slowed to 2.2% in 2014 and 0.9% in 2015 (NBS, 2015a, 2016a). Second, emissions growth fell to around 1%[3] in 2014, while in 2015 emissions appear to have actually declined in absolute terms by 0.1% (Korsbakken, Peters, & Andrew, 2016a). While considerable uncertainty surrounds official statistics (Grubb et al., 2015; Korsbakken, Peters, & Andrew, 2016b; Liu et al., 2015), 2014 may signal the start of a trend rather than an anomaly as China enters a 'New Normal' phase of economic growth

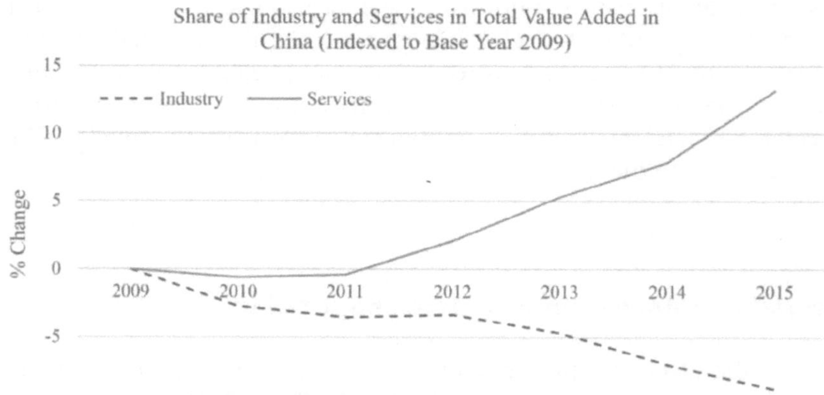

Figure 2 Structural shift from industry to services, 2009–2016
Source: Calculated from NBS (2016b) Online Database.

(Green & Stern, 2016; Song, Garnaut, Fang, & Johnston, 2015). Indeed these trends are expected to continue with increased focus on restructuring in China's 13th Five Year Plan, which includes still more ambitious targets for both the economy and environment (IMF, 2015; PRC, 2016).

An important consequence of increasing alignment between China's domestic economic and environmental agendas was to give China greater flexibility in its international climate policy, including more room for manoeuvre within international negotiations. Compared to Copenhagen, international climate agreements represented, by the end of the 12th FYP period, less of a threat to the economic progress on which domestic Chinese Communist Party legitimacy rests. In the 'New Normal' era of development, going low carbon, previously viewed as a burden that China was not yet willing to assume, began to be seen as a strategic opportunity that would allow China to address its growing domestic environmental crisis, mitigate the detrimental impact on China's international image of ever-increasing emissions, and build a strong position in the technologies that would be in increasing demand as the world accepted carbon constraints.

In practical terms, the slowdown in emissions resulting from a combination of slower, less energy-intensive growth and efforts to reduce the carbon intensity of the economy meant China was able, for the first time in 2014, to make an international commitment to peak emissions by 2030 – a goal now very much aligned with the 'New Normal' domestic policy agenda. This target is supported by a range of scenario-based studies, including by government agencies, some of which suggest it may be possible for emissions to peak as early as 2025 without major trade-offs in economic development (Chai & Xu, 2014; Energy Research Institute, Energy Foundation, 2015; He, 2014; Wang & Zou, 2014; Wang, Zhang, Pan, Yang, & Shi, 2014; Zhang, Karplus, Qi, Zhang, & He, 2016). By Paris, China was therefore in the enviable position of being able to propose a target that could potentially allow it to 'under-promise and over-deliver' (Green & Stern, 2016).

More importantly, China's ability to commit to a hard limit on emissions growth that was not imposed from outside removed a key barrier to cooperation with the US. This paved the way, in November 2014, for a landmark joint announcement on climate action between China and the US (White House, 2014). In addition to the commitment to peak emissions by 2030 or earlier, China announced an intention to increase the share of non-fossil energy to 20% by 2030. On the US side, President Obama reciprocated by announcing a 26–28% cut in emissions by 2025 on a 2005 baseline. The two sides also announced plans to strengthen cooperation on climate and clean energy through policy dialogue and technical cooperation, extended the mandate of the US–China Clean Energy Research Center to 2020, and renewed funding for building efficiency, clean vehicles, and advanced coal technologies with carbon capture, use, and sequestration. Other initiatives included a new joint initiative on the water-energy nexus, cooperation on hydrofluorocarbons, action on low-carbon cities, promotion of green goods trade, and clean energy pilot projects. Together, the US and China accounted for over one-third of global greenhouse gas emissions and the announcement was specifically aimed at injecting momentum into the global effort to decarbonize the world economy.

Cooperation deepened still further the following September with the US–China Joint Presidential Statement on Climate Change on the occasion of Xi Jinping's visit to the US (White House, 2015). The two leaders laid out a shared vision for a global climate agreement in Paris that prefigured the final result: it would aim to ramp up ambition over time and set the long-term goal of global low-carbon transformation. Specifically, they agreed on the need for enhanced transparency, reporting and review of action, and made progress on commitments to global climate finance that included a pledge from China of CNY 20 billion ($3.1 billion) in climate assistance to developing countries. They reaffirmed the principle of common but differentiated responsibilities and respective capabilities, in light of different national circumstances, but there were signs of a new flexibility in the interpretation of that principle, agreeing to embed differentiation in the relevant elements of the agreement as appropriate. They also agreed that more attention needed to be given to adaptation.

5. A more cooperative conference

As Paris approached, it seemed likely that China would be in support of a flexible, comprehensive, legally binding agreement. At the same time, success was far from guaranteed and there are a large number of factors beyond changes in China's domestic policy landscape that help explain why countries were able to achieve in 2015 what they could not in 2009. Detailed discussion of these factors is beyond the scope of this article (see Kinley, 2016), but it is worth noting at least two salient points.

First, impressive diplomatic efforts on the part of the French hosts, which began two years before the COP itself, are widely acknowledged to have played an important role in delivering the Paris Agreement. Led by COP chair and former foreign minister, Laurent Fabius, they included wide consultations with the major groups and important players – including China – to identify obstacles, understand domestic priorities, and to begin to align the interests of the parties to the common goal of reaching a comprehensive, but flexible agreement (Kinley, 2016).

Second, building on French diplomatic efforts, changes in the negotiations themselves helped lay the ground for a more cooperative rather than confrontational conference. In particular, the shift away from legally binding obligations to 'contributions', in the form of Intended Nationally Determined Contributions submitted through a system of 'pledge and review', helped transform proceedings from a zero-sum game into a more virtuous process of cooperation, if not an outright race-to-the-top (see also Rogelj et al., 2016). By addressing the issue of sovereignty, it also helped to allay long-held Chinese concerns about submitting to international MRV and diminish resistance in developing countries, including China, to moving away from the binary Annex system that had undermined progress in Copenhagen. Now all countries were encouraged to contribute to emissions reductions, further enhancing the collaborative spirit of negotiations

This is not to argue that there were no nervous moments. The Chinese delegation appeared to find the appearance of the 'High Ambition Coalition' (HAC) disconcerting. The combination of rich and poor countries in the HAC removed the last traces of the rigid division between Annex I and non-Annex I countries and undermined China's residual claim to represent the interests of the developing world (Christoff, 2016). Chinese negotiators also briefed against the 1.5°C target, complaining that it was 'unrealistic' and that there was no pathway to reach it (Hilton et al., 2015). That was probably true: the 1.5-degree goal was arguably a belated recognition of the perilous situation of the most vulnerable countries, and a negotiating card, not a plan. But there was no clear pathway to the 2-degree goal either, and inclusion of 1.5 degrees in the agreement acknowledged the gravity of the changes already underway in the climate system, even if it did not offer an immediate remedy.

6. Moving forward: from agreement to action

The Paris Agreement does not by itself put the world on a sustainable path to avoiding dangerous climate change (Christoff, 2016; Höhne et al., 2016). In particular, further efforts will be needed to follow through with transparent MRV and ensure that countries do not use the pledge-and-review system to backslide on earlier levels of ambition. In China, a range of challenges have the potential to hinder emissions reduction (Li, 2016). Economically, the transition to the 'New Normal' may not be smooth and risks potential resistance from vested interests in both state-owned enterprises and local governments impacted by the shift away from heavy industry. In the power sector, there are challenges associated with system planning and operation, establishing a credible regulatory system, moving towards market-based pricing, integration of renewables, and addressing coal over-capacity. Meanwhile continuing capital flight and high levels of debt could delay or even derail the execution of long-term plans. At a more fundamental level, it is also possible that China reduces domestic emissions but continues to export emissions overseas – either directly through building coal-fired power stations and infrastructure, or indirectly as energy-intensive industrial production increasingly moves from China to other parts of the world.

Despite these caveats, however, the value of the Paris Agreement is that it provides a clear and transparent framework for climate action involving all countries. It was made possible, at least in part, by increasing alignment between China's economic and environmental policy agendas. While considerable uncertainty exists around future emissions trajectories, it seems likely that China's shift towards low-carbon development – and a more positive stance in climate negotiations – is intended to be both structural and long term. Further evidence of this shift was provided in September 2016, when China became the first major emitter to ratify the Paris Agreement, followed shortly after by the US. This will help to keep open at least the possibility that the Paris Agreement's two-degree temperature goal can be met – an undertaking that will likely require China to become not just increasingly cooperative, but increasingly proactive on the international stage.

Disclosure statement

No potential conflict of interest was reported by the authors.

Notes

1. The periodization used here follows that of Green and Stern (2016) in their important contribution to understanding the environmental implications of China's changing economic structure.
2. Industry here refers to 'secondary industry' and services to 'tertiary industry' as defined by the National Bureau of Statistics. See All China Data Center for summary (http://chinadataonline.org/info/yearly03.asp).
3. Uncertainty surrounding emission factors and revisions of official figures on fuel consumption has led to discrepancy among estimates. Korsbakken et al. (2016b) put 2014 emissions growth at 0.8%, compared to 1.2% in Jackson et al. (2015).

References

AP (Associated Press). (2015). *China played a key role in global climate pact.* Retrieved July 2016, from www.sfgate.com/world/article/China-played-key-role-in-global-climate-pact-6695812.php

Bodansky, D. (2010). The Copenhagen climate change conference: A postmortem. *American Journal of International Law, 104*(2), 230–240.

Cai, F. (2012). Is there a "middle-income trap"? Theories, experiences and relevance to China. *China & World Economy, 20*(1), 49–61.

CCCPC. (2013). *Decision on major issues concerning comprehensively deepening reforms* [Chinese]. Retrieved July 2016, from http://news.xinhuanet.com/politics/2013-11/15/c_118164235.htm

Chai, Q. M., & Xu, H. Q. (2014). Modeling an emissions peak in China around 2030: Synergies or trade-offs between economy, energy and climate security. *Advances in Climate Change Research, 5*(4), 169–180.

Christoff, P. (2010). Cold climate in Copenhagen: China and the United States at COP15. *Environmental Politics, 19*(4), 637–656.

Christoff, P. (2016). The promissory note: COP 21 and the Paris climate agreement. *Environmental Politics, 25*(5), 1–24.

Clémençon, R. (2016). The two sides of the Paris climate agreement dismal failure or historic breakthrough? *The Journal of Environment & Development, 25*(1), 3–24.

Conrad, B. (2012). China in Copenhagen: Reconciling the "Beijing climate revolution" and the "Copenhagen climate obstinacy". *The China Quarterly, 210*, 435–455.

Energy Research Institute, Energy Foundation. (2015). *China 2050 high renewable energy penetration scenario and roadmap study.* Retrieved July 2016, from http://www.efchina.org/Attachments/Report/report-20150420/China-2050-High-Renewable-Energy-Penetration-Scenario-and-Roadmap-Study-Executive-Summary.pdf

Garnaut, R., & Huang, Y. (2005). The risks of investment-led growth. In R. Garnaut & L. Song (Eds.), *The China boom and its discontents* (pp. 1–18). Canberra, Australia: Asia Pacific Press.

Geall, S., & Hui, L. N. (2015). *China's low carbon future offers global opportunities.* China Dialogue. Retrieved July 2016, from https://www.chinadialogue.net/reports/8362-China-s-low-carbon-future-offers-global-opportunities/en

Green, F., & Stern, N. (2016). China's changing economy: Implications for its carbon dioxide emissions. *Climate Policy.* doi: 10.1080/14693062.2016.1156515.

Grubb, M., Sha, R., Spencer, T., Hughes, N., Zhang, Z. X., & Agnolucci, P. (2015). A review of Chinese CO_2 emission projections to 2030: The role of economic structure and policy. *Climate Policy, 15*(S1), S7–S39.

He, J. K. (2014). An analysis of China's CO_2 emission peaking target and pathways. *Advances in Climate Change Research, 5*(4), 155–161.

Hilton, I., Shuo, L., Zhang, Z., Lewis, J., Seligsohn, D., & Wang, A. (2015). *Is China a leader or a laggard on climate change?* Retrieved July 2016, from https://www.chinafile.com/conversation/china-leader-or-laggard-climate-change

Höhne, N., Kuramochi, T., Warnecke, C., Röser, F., Fekete, H., Hagemann, M., … Gonzales, S. (2016). The Paris agreement: Resolving the inconsistency between global goals and national contributions. *Climate Policy, 17*(1), 16–32. doi: 10.1080/14693062.2016.1218320.

IMF. (2015). *People's Republic of China: Staff report for the 2015 Article IV consultation.* Retrieved July 2016, from http://www.imf.org/external/pubs/ft/scr/2015/cr15234.pdf

Jackson, R. B., Canadell, J. G., Le Quéré, C., Andrew, R. M., Korsbakken, J. I., Peters, G. P., & Nakicenovic, N. (2015). Reaching peak emissions. *Nature Climate Change, 6*, 7–10.

Kinley, R. (2016). Climate change after Paris: From turning point to transformation. *Climate Policy, 17*(1), 9–15. doi: 10.1080/14693062.2016.1191009.

Korsbakken, J. I., Peters, G. P., & Andrew, R. M. (2016a). *China's coal consumption and CO_2 emissions: What do we really know?* Retrieved July 2016, from https://www.chinadialogue.net/article/show/single/en/8780-China-s-coal-consumption-and-CO2-emissions-What-do-we-really-know-

Korsbakken, J. I., Peters, G. P., & Andrew, R. M. (2016b). Uncertainties around reductions in China's coal use and CO2 emissions. *Nature Climate Change*, 6, 687–690.

Li, A. H. (2016). Hopes of limiting global warming? China and the Paris agreement on climate change. *China Perspectives*, (1), 49.

Li, J., & Wang, X. (2012). Energy and climate policy in China's twelfth five-year plan: A paradigm shift. *Energy Policy*, *41*, 519–528.

Liu, Z., Guan, D., Wei, W., Davis, S. J., Ciais, P., Bai, J., ... Andres, R. J. (2015). Reduced carbon emission estimates from fossil fuel combustion and cement production in China. *Nature*, *524*(7565), 335–338.

Lynas, M. (2009). How do I know China wrecked the Copenhagen deal? I was in the room. *The Guardian*, *22*(2009), 8.

NBS (National Bureau of Statistics). (2015a). *Statistical communique of the People's Republic of China on the 2014 national economic and social development*. Retrieved July 2016, from http://www.stats.gov.cn/english/PressRelease/201502/t20150228_687439.html

NBS (National Bureau of Statistics). (2015b). *The new normal: New strategy, new kind of development – the striking success of China's economic and social development in the "12th Five Year Plan"* [Chinese]. Retrieved July 2016, from http://www.stats.gov.cn/tjsj/zxfb/201510/t20151013_1255154.html

NBS (National Bureau of Statistics). (2016a). *Statistical communique of the People's Republic of China on the 2015 national economic and social development*. Retrieved July 2016, from http://www.stats.gov.cn/english/PressRelease/201602/t20160229_1324019.html

NBS (National Bureau of Statistics). (2016b). *Online statistics database*. Retrieved July 2016, from http://data.stats.gov.cn/english/index.htm

NDRC. (2015). *China medium and long term energy conservation plan*. Retrieved July 2016, from http://fourfact.com/images/uploads/China_Energy_Saving_Plan.pdf

PRC (People's Republic of China). (2011). *Twelfth five year plan (full text)* [Chinese]. Retrieved July 2016, from http://www.china.com.cn/policy/txt/2011-03/16/content_22156007_2.htm

PRC (People's Republic of China). (2016). *Thirteenth five year plan (full text)*. Retrieved July 2016, from http://www.china.com.cn/lianghui/news/2016-03/17/content_38053101.htm

Putnam, R. D. (1988). Diplomacy and domestic politics: The logic of two-level games. *International organization*, *42*(03), 427–460.

Qi, T., Winchester, N., Karplus, V. J., & Zhang, X. (2012). *Will economic restructuring in China reduce trade-embodied CO_2 emissions?* MIT Joint Program on the Science and Policy of Global Change.

Revkin, A. C., & Broder, J. M. (2009). A grudging accord in climate talks. *New York Times*, *19*.

Rogelj, J., den Elzen, M., Höhne, N., Fransen, T., Fekete, H., Winkler, H., ... Meinshausen, M. (2016). Paris agreement climate proposals need a boost to keep warming well below 2°C. *Nature*, *534*(7609), 631–639.

Song, L., Garnaut, R., Fang, C., & Johnston, L. (2015). *China's domestic transformation in a global context*. Acton, Australia: ANU Press.

State Council of the People's Republic of China. (2014). *Energy development strategic action plan (2014–2020)*. Retrieved July 2016, from http://www.gov.cn/zhengce/content/2014-11/19/content_9222.htm

Stern, N. (2011). Raising consumption, maintaining growth and reducing emissions. *World Economics*, *12*(4), 13–34.

UNFCCC and Boden, T. A., Marland, G., & Andres, R. J. (2015). *Global, regional, and national fossil-fuel CO_2 emissions*. Carbon Dioxide Information Analysis Center, Oak Ridge National Laboratory, U.S. Department of Energy, Oak Ridge, TN. doi:10.3334/CDIAC/00001_V2015

Wang, Y., & Zou, L. L. (2014). The economic impact of emission peaking control policies and China's sustainable development. *Advances in Climate Change Research*, *5*(4), 162–168.

Wang, Z. X., Zhang, J. J., Pan, L., Yang, F., & Shi, L. N. (2014). Estimate of China's energy carbon emissions peak and analysis on electric power carbon emissions. *Advances in Climate Change Research*, *5*(4), 181–188.

White House. (2014). *U.S.-China joint announcement on climate change.* Retrieved July 2016, from https://www.whitehouse.gov/the-press-office/2014/11/11/us-china-joint-announcement-climate-change

White House. (2015). *U.S.-China joint presidential statement on climate change.* Retrieved July 2016, from https://www.whitehouse.gov/the-press-office/2015/09/25/us-china-joint-presidential-statement-climate-change

Xinhua. (2015). *CPC Central Committee and State Council publish "ecological civilization master plan"* [Chinese]. Retrieved July 2016, from http://www.gov.cn/guowuyuan/2015-09/21/content_2936327.htm

Yan, M. (2015). How China can take the lead in adapting to the "new normal" economy [Chinese]. *China Central Government Website.* Retrieved July 2016, from http://www.gov.cn/zhengce/2015-08/13/content_2912124.htm

Ye, Q., & Wu, T. (2015). China's 'yes' to new role in climate battle. *Brookings Brief.* Retrieved July 2016, from www.brookings.edu/research/articles/2015/12/04-chinas-yes-new-role-climate-battle-qiwu

Zhang, X., Karplus, V. J., Qi, T., Zhang, D., & He, J. (2016). Carbon emissions in China: How far can new efforts bend the curve? *Energy Economics, 54*, 388–395.

Zhang, Z. (2015). Climate mitigation policy in China. *Climate Policy, 15*(S1), S1–S6.

Zhu, A., & Kotz, D. M. (2011). The dependence of China's economic growth on exports and investment. *Review of Radical Political Economics, 43*(1), 9–32.

Responsibility and liability for climate loss and damage after Paris

EMMA LEES

This article considers the loss and damage provisions of the Paris Agreement emerging from the 21st Conference of the Parties (COP) of December 2015, and examines the extent to which they represent an innovation from the voluntary-focused Warsaw International Mechanism (WIM). It concludes that whilst on its surface the Paris Agreement very much continues the WIM approach, the combination of the reference to differentiated responsibilities, the inclusion of a loss and damage article within the main Agreement, and the refusal to allow this article to act as a basis for liability raises some legal and policy ambiguities. These ambiguities raise the question as to whether it is necessary or useful to develop a responsibility allocation mechanism (which need not give rise to liability) in order to assess differentiated responsibilities for loss and damage, and if so, how such a mechanism might be constructed. This article then goes on to consider two alternative approaches to loss and damage: an insurance-based approach and an approach based on the twin pillars of causal connection and fault. The difficulties of designing an allocation mechanism on the basis of causation and fault are discussed. Nevertheless, it is concluded that if, in the future, as seems likely, there will be increasing pressure to determine some sort of responsibility allocation, these issues will need to be grappled with head-on.

Policy relevance

The article demonstrates the need to develop policy which is sensitive to the fine balance struck in the Paris Agreement between responsibility and liability for loss and damage, and prompts an open discussion as to how such responsibility ought to be allocated. It suggests that without such an open discussion, the effects of the Paris Agreement in relation to loss and damage will be limited.

1. Introduction

The Paris climate change negotiations at the 21st Conference of the Parties (COP) marked a significant change, as has been highlighted in this special issue, in the overall international approach to tackling climate change. The optimism which Paris engendered has led many to welcome its provisions, and the compromises, cooperation, and mutual support that these represent (e.g. see Kinley, 2016). However, there are some elements of the Paris Agreement which have been met with a greater degree of scepticism, and perhaps with some justification (e.g. see Sharma, 2016). Not least amongst these are the provisions within the Agreement, and its accompanying text, which relate to the allocation of responsibility for loss and damage arising from climate change. Legal responsibility and liability for such damage have represented a stumbling block to agreement in the past. The Paris Agreement's approach to compromising such tensions is, however, one which represents an unusual legal structure, obscuring both its policy significance and its legal consequences. Ultimately, it maintains the voluntary, cooperation-focused approach characterized by the Warsaw International Mechanism (WIM) established at COP 19 in Warsaw in 2013 (see Decisions 3/CP.18 and 2/CP.19). However, in this compromised structure lies the potential for progress.

The very existence of this structure demonstrates that the overall approach to questions of responsibility for loss and damage has shifted, at least in terms of attitude, and the potential for further steps in the direction of liability for climate change is clear. This article will demonstrate this by, firstly, considering the specific approach of the Paris Agreement to such loss and damage issues. It will then consider, briefly, the history of loss and damage provisions to provide some context to the Paris outcomes. It will then assess two aspects of the policy response required in the face of such unusual legal structures: firstly, it argues that there is a need to develop a response to the legal ambiguity, that is, the relevance of the distinction between responsibility and liability, to which these provisions give rise. It argues that this response must be clear and workable such that the ambiguity does not represent an excuse for inaction. Secondly, it argues that awareness of the challenges of assigning legal responsibility for loss and damage is critical to designing policy which is sensitive both to the outcomes of Paris and to the difficulties in any allocation of 'differentiated responsibility'.

In short, if we recognize that Paris distinguishes between responsibility and liability, we are liberated to consider how responsibility can be allocated without fear of immediate liability. Such a liberation brings with it challenges. It may, however, be necessary if insurance-based solutions are perceived as inadequate. This can be seen if we consider a potential mechanism for such allocation (causation and fault). Being aware of these challenges will improve our policy responses to Paris.

2. Articles 8 and 51

Before considering these specific points however, it is useful to outline what was decided at Paris in relation to loss and damage.

The most significant provision in relation to loss and damage is Article 8, which forms part of the Agreement itself. It states that:

> 1. Parties recognize the importance of averting, minimizing and addressing loss and damage associated with the adverse effects of climate change, including extreme weather events and slow onset events . . .
>
> . . .
>
> 3. Parties should enhance understanding, action and support, including through the Warsaw International Mechanism, as appropriate, on a cooperative and facilitative basis with respect to loss and damage associated with the adverse effects of climate change.
>
> 4. Accordingly, areas of cooperation and facilitation to enhance understanding, action and support may include:
>
> (a) Early warning systems;
>
> (b) Emergency preparedness;

(c) Slow onset events;

(d) Events that may involve irreversible and permanent loss and damage;

(e) Comprehensive risk assessment and management;

(f) Risk insurance facilities, climate risk pooling and other insurance solutions;

(g) Non-economic losses;

(h) Resilience of communities, livelihoods and ecosystems.

...

It is then confirmed in paragraph 51 of the decision to adopt the Agreement, however, that Article 8 does not give rise to liability. Thus, paragraph 51 states that: '... Article 8 of the Agreement does not involve or provide a basis for any liability or compensation'. Nevertheless, in the text of Article 8 are significant indications that the parties to the Agreement at least hope that the voluntary attitude to the WIM will extend further such that 'action and support' are facilitated, albeit on a 'cooperative' basis.

The interaction between Article 8 and paragraph 51, the precise terminology agreed, and the policy mechanisms which may emerge as a result will all be critical to the success of the loss and damage approach which the Paris Agreement represents. However, whilst the Paris Agreement is fundamentally premised on the concept of differentiated responsibilities, it does not, in relation to loss and damage, provide for tools for assessing these divergent responsibilities for fear of providing a foundation for liability. Whilst this failure is apparent throughout the Agreement (there is similarly no attempt to allocate responsibility in relation to mitigation and adaptation, for example), it is particularly significant in the context of loss and damage.

3. A point about terminology

Before analysing the issues raised by these provisions, it is necessary to explain the terminology being employed here. Firstly, in relation to loss and damage, the definition of such key terms is itself contested (Burkett, 2014, p. 120; UNFCCC Subsidiary Body for Implementation, 2012, p. 4; Wallimann-Helmer, 2015, p. 470). As can be seen, the Paris Agreement provides no firm definition of this concept and indeed this ambiguity may itself be a (deliberate) means of limiting the practical effects of Article 8. Rather, the Agreement provides some examples of loss and damage, including 'extreme weather events' and 'slow onset events'. The main definition in the text appears to be, however, that loss and damage are 'associated with the adverse effects of climate change'. This definition is very broad, and does not provide any clear guidance as to what constitutes loss and damage *per se* (as opposed to, for example, costs arising through the adoption of mitigation or adaption measures, both of which could be labelled as 'adverse effects' (Wallimann-Helmer, 2015, p. 471)).

This poses challenges in terms of assessing exactly what the consequences of including an article such as Article 8 will be. However, there has already been extensive consideration within the literature as to precisely how to define such key concepts. The reference in Article 8 to the WIM too gives some assistance in defining loss and damage. Primarily, the debate in the literature, and in practice, has centred around the extent to which loss and damage can, and should be, distinguished from adaptation and mitigation approaches (Dow et al., 2013; Warner & Van Der Geest, 2013). A consensus has emerged which highlights that however much is achieved through mitigation and adaption, some residual damage will still occur (Wallimann-Helmer, 2015, p. 472; Wrathall et al., 2015, p. 277). It is this which is referred to as loss and damage.

However, it is important to recognize, as Wallimann-Helmer (2015) has done, that since the 'driver' for adaptation is avoiding loss and damage, the boundary line between the two is not, and perhaps cannot be, clear. An example of this could be coastal protection measures resulting in the loss of coastal habitats. The measures are taken as a response to climate change, and are therefore rightly considered as adaptation, but the loss of the coastal habitat is permanent. This terminological difference is critical however, because any sensible response to questions of loss and damage must be able to distinguish these from adaptation measures. Furthermore, the distinction forms part of the toolbox for assigning responsibility for the various outcomes and challenges posed by climate change, and as such, the definition provided must be resilient in the face of political or evidential difficulties and predictable in its application. Robust definitions of loss and damage are, therefore, essential to the success of any policy seeking to consider collective responsibility for such outcomes. Such a definition is not present in the text of the Paris Agreement itself and nor will it ever be straightforward to apply. The terminological difficulty in distinguishing between loss and damage and adaptation is itself a stumbling block to progress, therefore, in this new field of enquiry.

Understanding that the Paris Agreement does not necessarily adopt such a definition, however (and as a result, aware that even with a clear definition of responsibility, the potential for obfuscation as to 'responsibility for what' remains), but recognizing the need to rely on a consistent and predictable definition, this article follows Burkett in concluding that 'loss refers to climate-related impacts for which restoration is not possible' (2014, p. 120) and that 'damage refers to negative impacts for which restoration is possible' (2014, p. 121). The key to these definitions is that, for both, the negative impacts have in fact taken place, whereas in adaptation situations, costs are incurred in attempting to avoid or divert such negative impacts. Thus, as Walliman-Helmer argues, the distinction between the two is as much a matter of goal (i.e. what is the measure trying to achieve prevention, restoration, or compensation for loss) as it is of practical effects (2015, p. 473). The line, inevitably, is a fine one, but is essential to the proper operation of any system which attempts to allocate responsibility for loss or damage, or indeed one which simply seeks to encourage a sharing of such responsibility, as the Paris Agreement does.

The second terminological clarification required relates to the meaning of the terms responsibility and liability. Article 8 of the Paris Agreement does not itself mention either responsibility or liability. However, the preamble to the Paris Agreement, unsurprisingly, refers to 'the principle of equity and common but differentiated responsibilities and respective capabilities, in the light of different national circumstances'. Decision 3/CP.19 which forms the basis of the WIM too refers to the need to take account of differentiated responsibility for loss and damage. The differentiated responsibility aspect of global climate change agreements is crucial to considering how a loss and damage approach may work in practice, since, to some extent, it must be considered that responses to loss and damage require recognition of responsibility. How, then, can responsibility be defined?

This, again, is a controversial question. However, it is important to note that there is a distinction to be drawn between responsibility (that is to say, a link between actor and factual outcome, be that a causal or other link as defined by the legal system (Honoré, 2010)) and liability (i.e. the legal consequence or burden which befalls them (Honoré, 2010; Luizzi, 1983). To put this at its most basic: responsibility is the relevant factual connection between the act and the 'real-world' outcome, and liability is a legal or other normative consequence of finding that such a connection exists whereby the connection is treated as an *a priori* justification for that consequence. A finding of responsibility is not, however, necessarily followed by a conclusion that liability is justified. Thus, 'to establish responsibility it must be shown that the harm was done ... by the agency that the law treats as a *potential* basis for liability' (Honoré, 2010)) (emphasis added).

Responsibility for an act is insufficient to give rise to legal consequences, although a finding of responsibility is a legal conclusion. Responsibility is, in this sense, a non-normative question. The law must also conclude that such responsibility justifies some consequence in order for liability to arise. The phraseology used in the Paris Agreement itself, deliberately distinguishing between responsibility in the preamble and liability in the accompanying decision, demonstrates that the parties to the Agreement recognize that a distinction between the two is possible and useful.

In climate change negotiations over the years, both the questions of responsibility and liability have been taboo, as Doelle (2014) notes, but they are, as has been shown, separable, and there is no immediate reason why allocation of responsibility for climate change loss and damage (relying, as it would, on scientific evidence demonstrating a causal link between action and outcome – a non-normative assessment (Fumerton & Kress, 2001, p. 84)) would necessarily produce liability of the type envisaged in paragraph 51, that is, a direct claim for compensation or other form of liability. It is therefore important to understand the difference between the two, and therefore precisely what paragraph 51 excludes.

Paragraph 51 excludes direct claims for *compensation*, or other kinds of liability, arising from loss or damage *responsibility*. Importantly, other forms of liability in this context could include a requirement to contribute a certain level to a collective pot (rather than a loss-related compensation claim) and as such, the prohibition on liability arising for loss and damage would exclude a responsibility-based mandatory requirement to contribute to an insurance pool (and as such, not all insurance solutions avoid the responsibility allocation question – see below). It is not, however, a barrier to increased attempts to allocate responsibility for loss and damage *per se*, and indeed, other areas of the text recognize the need to examine the collective balance of such responsibility.

By highlighting the distinction between responsibility and liability, it is possible to develop a nuanced understanding of the Paris Agreement which does not prevent the development of a responsibility allocation mechanism, as long as such a mechanism does not represent a basis for liability. There are many reasons why it is critical to examine questions of responsibility, without the outcome being one simply of liability, not least because it allows for a better understanding of the differentiated responsibilities to which the text refers, important given Friman and Hjerpe's conclusion, for example, that, 'knowledge is still scarce regarding how climate change policy makers understand historical responsibility' (2015, p. 303).

It may be argued that any attempt to assign responsibility inevitably results in some form of liability, but this is simply not the case. Assigning responsibility is useful in both moral and practical terms, but there is no need to assign any form of liability to such a conclusion. Nor is it is correct to argue that differentiating responsibility will inevitably result in the eventual imposition of liability. Liability is a legal consequence of a finding of responsibility. The best way to ensure that liability did not result from such a finding would be to assign a different consequence, for example, the imposition of a non-enforceable or moral duty to support or provide finance to those affected by loss and damage. The importance of this conclusion is discussed in further detail below.

4. The history of loss and damage responsibility and liability

Before considering this further, however, it is necessary to examine how the Paris Agreement modifies our existing approach to such responsibility allocation. As Friman and Hjerpe (2015) have noted, allocation of responsibility for climate change poses a number of significant problems: interpretative problems (what is loss and damage?); problems of justification for liability and, relatedly, the issue of retrospective liability provisions (how can we justify liability even once responsibility is determined?); and finally, problems of causal allocation and proof. These underlying difficulties have meant that agreement relating to loss and damage responsibility and liability has been hard to achieve to date, and explains in part the history of such liability which has been heavily focused on voluntary or symbolic approaches.

In part too this is due to the reluctance of developed nations to accept that they should bear the burden of consequences for historical sources of climate change (Friman & Hjerpe, 2015, p. 314). This reluctance, however, is arguably more of a question of fear of liability than it is an acceptance of some causal responsibility. Nevertheless, the result is that any agreements relating to loss and damage have been very modest in their ambitions. Nowhere is this clearer than in relation to the WIM, the most significant step in the historical development of global responses to loss and damage prior to the Paris Agreement.

This mechanism, which is to be revisited at COP 22 in 2016 but which is given full support in the Paris Agreement in Article 8, is primarily concerned with building knowledge, ensuring cooperation, and facilitating the sharing of expertise. Thus, the mechanism is designed to facilitate the enacting of Decision 3/CP.18 whose preamble states that it is based on, 'the need to strengthen international cooperation and expertise in order to understand and reduce loss and damage associated with the adverse effects of climate change'. Nowhere is there mention of allocating responsibility or liability for such effects. Rather, the response to the effects is, it is hoped, to be shared, not left to the 'victims' of such loss and damage. However, this sharing (through, in part, the concept of differentiated responsibility) is to be carried out on the basis of mutual action and cooperation, not through enforceable liability.

Again, paragraph six of this decision:

> Invites all Parties, taking into account common but differentiated responsibilities and respective capabilities and specific national and regional development priorities, objectives and circumstances, to enhance action on addressing loss and damage associated with the adverse effects of climate change.

It does not itself consider the mechanism by which responsibility is differentiated, nor does it engage with the significant policy challenges of so-doing. Rather, the question of differentiated responsibility is left, as with the definition of loss and damage, to mutually cooperative and vague reliance on concepts such as proportionality. Thus, to date at least, the approach to loss and damage has been very much based upon voluntariness following recognition of the need for mutual support with some moral responsibility derived from proportional contribution to historical emissions. This approach skirts around the need to assign responsibility by attaching no consequences to responsibility. The historical development of loss and damage mechanisms is, therefore, reflected in the Paris text. What the Agreement may provide however, adding to the historical developments, is a further impetus to examine mechanisms for assigning responsibility (even if they do not give rise to liability), and this will assist in the sharing of goals of the WIM as well.

5. Policy responses to legal ambiguity

Thus, it is clear that the approaches which currently exist to responsibility and liability for loss and damage arising from climate change, by avoiding the question of responsibility *per se*, cannot get onto the question of liability (i.e. the consequence imposed given the existence of responsibility). Therefore, the paragraph 51 denial of liability and compensation fits entirely with a failure to commit to any one understanding of or approach to assigning responsibility. Similarly, the decision not to provide a clear, predictable definition of loss and damage can also be said to limit the practical consequences of the acceptance of loss and damage responsibility.

Crucially, the consequences of these failures for our future policy approaches to funding and supporting those subject to loss and damage arising through climate change may be significant. The ambiguity may act as an excuse for inaction. It is important, therefore, to examine some potential responses to the failure to commit to a mechanism for allocation so as to avoid such inaction. This article will consider two potential responses: insurance, and responsibility allocation. It considers allocation of responsibility on the basis of causation and fault (there are, of course, other potential ways to allocate responsibility, not least strict liability such that a causal link alone would give rise to responsibility).

History favours the insurance mechanism. It allows parties and policy makers to avoid the question of allocation of responsibility all together. As such, it could be labelled a risk transfer system in that it spreads the 'financing' of remediation for loss throughout various parties, rather than confining them to the victim. It is this approach which was discussed at length in earlier COP negotiations, and it is in insurance approaches that many academics see the most potential for sharing of costs of loss and damage (Warner et al., 2012). Insurance mechanisms, from a legal perspective, are useful, in that they allow for the provision of funds without ever needing to specifically assign responsibility for the event which has taken place. For example, states could be required to contribute to an insurance pool on the basis of their gross domestic product or similar, rather than relying on their previous emissions on the basis of historical responsibility, or their ongoing emitting activities.

Consider here a car accident. The pedestrian has been consuming alcohol and stumbles into the road and is then hit by the driver exceeding the speed limit. Both are causally responsible for the accident. How do we decide, in such a scenario, who should face the cost of the pedestrian's medical care? Let us say that the pedestrian did not know that he or she was consuming alcohol, and the driver, whose speedometer was faulty, did not know that he or she was speeding. At this point, who should cover the costs? Alternatively, let us imagine that the pedestrian is aware that he or she has consumed excessive alcohol, but the driver is unaware of the speed. In such a case, how might we allocate costs liability then? The parallels to climate change are obvious (for discussion of the difficulties in reliably attributing factual cause in relation to climate change, see Huggel, Stone, Auffhammer, & Hansen, 2013;

Wrathall et al., 2015). In particular, developed countries, up until a certain point in the relatively recent past, were unaware that their consumption was producing climate change effects (Wrathall et al., 2015, p. 279), whereas those countries which are developing now are so aware. Any attempt to allocate responsibility runs into these issues, and so any approach which requires some sense of causation and fault is difficult to develop (some indications as to how such might be achieved are discussed below).

It is precisely because of these difficulties that insurance approaches have become popular within the literature. They do not depend upon causation and fault, since insurance requires collective pooling of resources, and the allocation of funds depends upon the need of the 'victim'. It is for this reason – its ability to bypass the question of responsibility – that insurance mechanisms have been the norm in relation to climate change negotiations for loss and damage. However, the Paris Agreement itself, and its decision text in paragraph 48, envisage insurance and risk transfer mechanisms as being only one of a series of possible responses to the loss and damage question. This signals a shift away from an approach where 'loss and damage' is seen as a collective problem, solved through solely collective solutions, to one where the question of responsibility raises its head. The Paris Agreement may not go so far as to specify how responsibility is assigned, but it does highlight that insurance-type responses may be inadequate, on their own, in solving such problems. It therefore suggests that we move beyond insurance or at least explore alternative options.

Thus, although no definition of loss or damage is given in the Agreement, and no mechanism is cited for the allocation of 'differentiated responsibility' for climate change (and, explicitly, no liability is to be associated with the loss and damage provisions), the Paris Agreement implicitly challenges us to consider mechanisms beyond insurance which would involve allocating responsibility for loss and damage (through causal mechanisms). The ambiguities in the Paris Agreement can, in short, be seen as deliberate obfuscation designed to prevent further debate as to responsibility allocation mechanisms, or they can be seen as the start of a genuine debate in relation to such mechanisms. The only consequence which is off the table is liability, and liability, although critical to loss and damage questions, is a possible, but not an inevitable, result of beginning to think about responsibility. This article suggests that the most appropriate response to this ambiguity is not to use it as a reason to sit idle, but rather indeed as a prompt to begin to consider responsibility more openly than has been the case to date in relation specifically to loss and damage.

6. Policy challenges of assigning legal responsibility for loss and damage

Thus, on the basis of the Paris Agreement text, two issues emerge. Firstly, the mechanism by which responsibility is allocated or 'differentiated', and in particular, the legal elements which would need to form a part of that allocation if it is to be successful, must be examined. The second issue is a forward-thinking one: assuming that such a nuanced and appropriate responsibility allocation mechanism can indeed be developed, how would this relate to liability and compensation if we so desired? These two issues will be addressed in turn.

First, it is useful to consider responsibility allocation mechanisms, so that we are better equipped to understand the concept of differentiated responsibility, and how this relates to loss and damage. There are many different potential mechanisms for so-doing. They could involve consideration of cause; of fault (in the sense of a certain type of knowledge, for example); of total emissions in terms of 'market share'; or of some other fact-based link between emission and emitter. In this article, as highlighted above, the type of responsibility discussed here involves the existence of a causal link, associated with some degree of culpability or fault, for a damaging action between an actor and an action. Two elements must, therefore, be defined in such a model: the causal link and the fault requirement. These two elements are at the core of most national legal systems, and as such a causation and fault-based concept of responsibility represents the most 'standard' (although by no means the only) response to questions of allocation, hence its consideration here.

Seeing climate change through the lens of loss and damage highlights the importance of engaging national, as well as international law in the allocation of responsibility, since it is here that experience in handling such questions lies. Indeed, Wrathall et al. (2015) recognize the immense difficulty that any attempt to regulate climate change responsibility at an international level engenders:

> due to the complex relationship between industrialisation, the production of GHGs and climate impacts, the establishment of a judicial body on climate change is neither practical, nor is this model politically feasible in UNFCCC or international decision-making bodies in general. (p. 286)

Harnessing the power of national legal systems or concepts developed at such a level may, therefore, be useful. However, this perspective on the question of loss and damage also highlights the dangers of engaging such national legal systems. Although they are able to compensate for the enforcement gap from which international law famously suffers, and can overcome some of the political difficulties outlined above, they will each be idiosyncratic in their approach to definitions of cause and fault, and any attempt to harmonize these approaches will be doomed to fail unless there is an overarching adjudication system. The first lesson to take from any attempt to frame responsibility for loss and damage arising from climate change through the lenses of causation and fault, therefore, is that there is a need not only to establish what decision needs to be made in terms of proving cause and fault, but also to consider who ought to make that decision.

This is not the only difficulty in relying on causation and fault, however, as a responsibility allocation mechanism. These terms, in themselves, are very difficult to define. This can be seen by focusing on the causal link between an action and an outcome. Similar issues arise in relation to fault. It is useful here to give the example of causation. The analysis here simply highlights the difficulties which emerge when this approach is taken to climate change. It does not suggest that these are the only ways to conceive of causation.

For most situations which a legal system is required to address, a 'but for' causal test will suffice: 'but for' the actions of person A, the event would not have taken place (Hart & Honoré, 1985). Thus, but for the speeding car above, the car accident would not have taken place. On this model, the car driver is causally responsible for the accident. So too, incidentally, is the drunken pedestrian. But in many cases involving climate change, the 'but for' test will be unhelpful. This would be the case, for example, if we consider CO_2 emissions. 'But for' the emissions of the UK, or Shell, or Mrs Smith's car, would this particular severe weather event have taken place? The answer is probably yes it would, but it would not have taken place were it not for the collective emission. The 'but for' test is, therefore, generally unhelpful in cases of cumulative damage.

If we do wish to consider causation as relevant to the approach to differentiating and allocating responsibility therefore, it is necessary to go beyond the 'but for' approach. What other approaches are there? Another way of thinking about cause is to look for sufficient elements within a necessary subset. Thus, assuming the UK emitted *x* tonnes of CO_2, and it is established through scientific inquiry that *x* tonnes of CO_2 *alone* would be sufficient to have caused a particular extreme weather event, whilst we cannot be sure that it was in fact the 'UK's tonnes' that caused the event, we can be sure that their actions were enough to cause the event. The UK's actions would be a sufficient part of a necessary subset, that is, the subset of CO_2 emissions.

A third option is the necessary element of the sufficient subset approach (Fumerton & Kress, 2001; Wright, 1985, 2011). Thus, it could be said that scientific evidence could not show whether CO_2 or methane emissions resulted in a particular event. However, it could be said that *x* tonnes of CO_2 *would* be enough to cause the event, whether or not CO_2 caused it or methane caused it (or both). Let us say then that in total *x* tonnes of CO_2 have been emitted. In such a case, *any* emitter that contributes to the *x* tonnes has contributed a necessary element (some tonnes) to a sufficient subset (the CO_2). Similarly, if *y* tonnes of methane would also have been enough on its own, then any emitter that has contributed to the *y* tonnes of methane has also contributed to the necessary subset. The problem with this approach emerges when the total amount of CO_2 or methane is $x + y + z$ tonnes. How do we decide whether the emitter's tonnes fell into *x* or *y* (the necessary element) or *z* ('surplus to requirements')? And if this has been decided, what difference does this make? By delving into the detail of causal mechanisms, it is possible to see the challenges posed by attempting to rely on causation as the basis for responsibility (even if no liability thereby emerges).

Thus, '[i]n order to ensure an equitable distribution of efforts, [allocation must be] based on agreed upon principles of equity resulting in an allocation formula' (Doelle, 2014, p. 40), and these principles would have to go beyond causal mechanisms, since focusing on cause alone does not tell us which of the three approaches outlined above – 'but for', sufficient element of a necessary subset, or necessary element of a sufficient subset – is the best solution to the allocation problem. As has been noted, similar problems would arise were we to look to fault, not least because it is difficult to say with certainty at what point in our recent history suspicion of climate change transformed into knowledge, such that emitters post this point would be at fault. Furthermore, a fault-based approach, by requiring knowledge, might in fact allocate responsibility more fully onto those countries currently developing, since they are aware of the damage being caused, rather than currently developed countries. This is not an intuitively 'correct' solution to the allocation of liability problem, although its precise operation would of course depend upon the date at which we assume that knowledge of climate change became widespread, and current, continuing rates of emission (perhaps on a *per capita* basis).

Assuming, however, that such an approach can be developed, the final issue to consider is the effect this would likely have in terms of compensation and liability. Whilst paragraph 51 explicitly excludes this, it is clear from the rhetoric surrounding the conference, not least from the 'victim' states, that liability, ultimately, may be necessary if sufficient support is to be provided to such states to allow them to adequately handle the loss and damage that they will suffer. Such liability, however, need not be limited to states themselves. Rather, once a responsibility allocation mechanism is developed, the potential for states to utilize this mechanism to transfer their own responsibilities onto private actors is clear. This is an outcome which many would welcome: but it is also one which should only be adopted openly, and with foresight. The danger, therefore, of developing an allocation mechanism is that it immediately transfers the eventual question of liability from states onto private actors, thus allowing states, to an extent, to duck out of their own role in climate change.

7. Conclusions

Despite these risks, however, it will still be necessary to begin down the path of the development of an allocation mechanism, most likely on the basis of cause and fault, if progress is to be made beyond the voluntary or insurance-focused approach into a more 'support'-focused approach as advocated to a certain extent by the Warsaw International Mechanism and as 'strengthened' by the Paris Agreement. The refusal to contemplate liability in the text is not a refusal to contemplate responsibility allocation, and indeed, the foundational concept of differentiated responsibility demands that steps be taken to consider this previously 'taboo' subject more openly. Wrathall et al. (2015) issue a challenge: 'Have we been creative enough in imagining policy alternatives' (p. 288) to the loss and damage problem?

It can be hoped that the new steps taken in Paris, and the optimism following the Agreement, prompt this open consideration and imagination, rather than allowing a retreat behind ambiguity. If such a consideration does take place, however, it will likely become necessary to grapple with questions of causation and fault, and as the discussion here of causation has shown, such issues are far from straightforward. Furthermore, if this path is taken, the implications for private actors become clear: once we have an allocation mechanism in place, it may be obvious that the real 'source' of the problem is not the State itself. Such an outcome may be far from unwanted, but it must be adopted openly. Thus, the Paris Agreement demands two policy responses: firstly, it advocates further consideration of responsibility allocation and specifically in relation to loss and damage issues; secondly, it demands that we recognize the consequences of such an allocation. This article has suggested some of the difficulties which may arise in so-doing. These difficulties should not, however, become reasons for inaction: they simply demand that the action which takes place is sensitive and open.

The Paris Agreement implicitly highlights the distinction between liability and responsibility, but it fails both to define loss and damage, and to develop a responsibility allocation mechanism. However, to make progress in this area, we should not be afraid to adopt such a mechanism since liability is not an inevitable consequence of responsibility. Indeed, adopting such a mechanism will be necessary if we are to supplement the current insurance-focused and voluntary approach, which may be needed as the losses from climate change become more serious and apparent. By starting to consider such allocation mechanisms, the usual mechanisms of fault and causation appear ill-equipped to handle this task in relation to climate change. It is important that we are aware of these challenges, and of the need to develop principles of equity which can help determine how we meet these difficulties. Only then can our policy response to the Paris Agreement be resilient in the face of the evidential and political barriers to which climate change gives rise.

Disclosure statement

No potential conflict of interest was reported by the author.

References

Burkett, M. (2014). Loss and damage. *Climate Law, 4*, 119–130.
Doelle, M. (2014). The birth of the Warsaw loss & damage mechanism. *Carbon and Climate Law Review, 8*, 35–45.
Dow, K., Berkhout, F., Preston, B. L., Klein, R. J. T., Midgley, G., & Shaw, M. R. (2013). Limits to adaptation. *Nature Climate Change, 3*, 305–307.
Friman, M., & Hjerpe, M. (2015). Agreement, significance, and understandings of historical responsibility in climate change negotiations. *Climate Policy, 15*, 302–320.
Fumerton, R., & Kress, K. (2001). Causation and the law: Preemption, lawful sufficiency, and casual sufficiency. *Law and Contemporary Problems, 64*, 83–105.
Hart, H. L. A., & Honoré, T. (1985). *Causation in the law* (2nd ed.). Oxford: Clarendon.
Honoré, T. (2010, Winter). Causation in the law. In E. N. Zalta (Ed.), *The Stanford encyclopedia of philosophy*. Retrieved February 20, 2016, from http://plato.stanford.edu/archives/win2010/entries/causation-law/
Huggel, G., Stone, D., Auffhammer, M., & Hansen, G. (2013). Loss and damage attribution. *Nature Climate Change, 3*, 694–696.
Kinley, R. (2016). Climate change after Paris: From turning point to transformation. *Climate Policy, 17*(1), 9–15. doi:10.1080/14693062.2016.1191009.
Luizzi, V. (1983). The bounds of liability and responsibility. *Journal of Social Philopsophy, 14*, 21–33.
Sharma, A. (2016). Precaution and post-caution in the Paris Agreement: adaptation, loss and damage, and finance. *Climate Policy, 17*(1), 33–47. doi:10.1080/14693062.2016.1213697
UNFCCC (United Nations Framework Convention on Climate Change) Subsidiary Body for Implementation. (2012, November 26–December 1, 2012). *A literature review on the topics in the context of thematic area 2 of the work programme on loss and damage: A range of approaches to address loss and damage associated with the adverse effects of climate change*. Thirty-Seventh Session, Doha.
Wallimann-Helmer, I. (2015). Justice for climate loss and damage. *Climatic Change, 133*, 469–480.
Warner, K., Kreft, S., Zissener, M., Höppe, P., Bals, C., Loster, T. ... Oxley, A. (2012). *Insurance solutions in the context of climate change-related loss and damage: Needs, gaps, and roles of the Convention in addressing loss and damage*. Submitted to the SBI Work Program on Loss and Damage. Retrieved February 20, 2016, from http://unfccc.int/resource/docs/2012/smsn/ngo/276.pdf
Warner, K., & Van Der Geest, K. V. (2013). Loss and damage from climate change: Local-level evidence from nine vulnerable countries. *International Journal of Global Warming, 5*, 367–386.
Wrathall, D. J., Smith, A. O., Fekete, A., Gencer, E., Reyes, M. L., & Sakdapolrak, P. (2015). Problematising loss and damage. *International Journal of Global Warming, 8*, 274–294.
Wright, R. W. (1985). Causation in Tort law. *California Law Review, 73*, 1735–1828.
Wright, R. W. (2011). The NESS account of natural causation: A response to criticisms. In R. Goldberg (Ed.), *Perspectives on causation* (pp. 285–332). Oxford: Hart.

Small group, big impact: how AILAC helped shape the Paris Agreement

GUY EDWARDS, ISABEL CAVELIER ADARVE, MARÍA CAMILA BUSTOS, AND J. TIMMONS ROBERTS

What role can one group of small and medium-sized countries play in breaking the long-standing impasse on climate change? What explains the formation of such a group and how can we assess its impact on outcomes such as the 2015 Paris Agreement? This article assesses the Independent Association of Latin America and the Caribbean (La Asociación Independiente de América Latina y el Caribe – AILAC) and its contribution towards cutting the Gordian Knot of climate interests and the building of a universal global regime on the issue. We review the origins and evolution of AILAC as well as its contributions to the Paris Agreement adopted in 2015 on five key issues. We conclude with an assessment on the group's future prospects and challenges.

Policy relevance

Understanding AILAC's origins, evolution, and contributions to the Paris Agreement has important implications for analysing the future possibilities and direction of global climate policy. AILAC represents a unique example of a group of small and medium-sized countries that succeeded in having an impact upon the pivotal 2015 Paris negotiations. Rather than originating from national leaders, the formation of the group was orchestrated organically by country negotiators, as they looked to increase their countries' visibility and influence at the negotiations. The article discusses the areas of the Paris Agreement where the group had an impact: *differentiation* between developed and developing country obligations, the *legal architecture* of the agreement, the format and review of the *Intended Nationally Determined Contributions* (INDCs), *climate finance*, and *adaptation* to climate impacts. Beyond the 21st Conference of the Parties, AILAC faces two crucial tests. First, whether the group's high ambition rhetoric at the United Nations Framework Convention on Climate Change (UNFCCC) can be matched by national policy advances on climate change and the implementation of the Nationally Determined Contributions. Second, can the group successfully consolidate its positions at the UNFCCC and overcome institutional challenges, while taking on new members? The AILAC case offers a significant example of norms and ideas effectively spreading between countries at the UNFCCC, while contributing to enhanced global and national action on climate change.

1. Introduction

Scarcely mentioned in the literature on global climate governance, the Independent Association of Latin America and the Caribbean (AILAC) is a group of eight countries participating in the negotiations under the United Nations Framework Convention on Climate Change (UNFCCC). AILAC was launched at the UN climate change conference in Doha, Qatar, in 2012 by Chile, Colombia, Costa

Rica, Guatemala, Panama, and Peru. In 2015, Paraguay and Honduras joined. AILAC is also part of the Group of 77 and China (G77) made up of 133 developing countries.[1] It focuses on taking a pragmatic approach to the UNFCCC negotiations, striving to generate coordinated and ambitious contributions to the negotiations, while following a vision for sustainable development that sees itself as responsible for the environment and future generations (AILAC website, accessed June 2016).

Before its creation, the group's members began to coordinate their positions at the UNFCCC in 2005, and have presented joint statements since 2007 (Friedman, 2013). Most AILAC members are also part of the informal group, the Cartagena Dialogue for Progressive Action, a space formed in 2010 where developed and developing countries share views on the negotiations and search for common positions.

Despite being only middle-income countries, AILAC has been one of the most visible negotiating groups at the UNFCCC negotiations since 2013. AILAC countries are highly vulnerable to climate change impacts, which partially explains why they promote increased ambition for reducing emissions and a stronger focus on adaptation. When looking at total greenhouse gas emissions (excluding land-use change and forestry), AILAC members had per capita emissions below the global average of 6.20 tCO_2e, varying from Guatemala's 1.54 tCO_2e to Chile's 5.77 tCO_2e.[2] While its positions have been simultaneously applauded and critiqued by different countries, the group has been dubbed a 'vital linchpin between progressive developing and developed countries'.[3] On the one hand, AILAC has been celebrated for its efforts to attempt to build bridges between countries.[4] On the other, the group has been labelled as a sell-out by some developing nations for its willingness to work with developed countries at the UNFCCC.[5]

AILAC countries are for the most part market-oriented, with economies relatively open to free trade and largely centrist – either centre-left or centre-right – governments.[6] Alongside other Latin American nations, AILAC countries are demonstrating a foreign policy that is increasingly independent of the US, reflected in the growing presence of new trade and diplomatic partners. AILAC promotes the idea that development can be decoupled from emissions, emphasizing the economic benefits of climate action (Garbiladi, 2014). The group's vision entails a greater emphasis on the diversification of national economies away from a focus on natural resource extraction and export and towards greater innovation and technologically advanced economies. However, given that a number of AILAC countries have a substantial reliance on the extraction of natural resources, achieving this vision remains a challenge (Edwards & Roberts, 2015).

AILAC countries have promoted ambitious positions at the UNFCCC negotiations, seeking to contribute to a high ambition global regime that can help trigger national policy developments. Central to AILAC's perspective is that domestic action can boost ambition at the multilateral level, which in turn can reinforce political will domestically in favour of decarbonization (Friedman, 2013). This article assesses AILAC's contribution to the UNFCCC negotiations, whose effectiveness has been repeatedly undermined by acute and chronic differences between countries from the Global North and Global South, and within these groups as well (Roberts & Parks, 2007). It reviews the origins and evolution of AILAC and evaluates its contribution to the 2015 Paris Agreement, specifically on differentiation between the expectations of developed and developing countries, the legal architecture of the new agreement and the Intended Nationally Determined Contributions (INDCs), mitigation, finance, and adaptation. It explores AILAC's future prospects and looks at whether its high ambition rhetoric at the UNFCCC can be matched by national climate policy advances and the implementation of the

Paris Agreement, and if the group can consolidate its positions at the UNFCCC vis-à-vis institutional challenges.

2. The origins and evolution of AILAC

The origins and evolution of AILAC came after two decades of slow progress at the UNFCCC negotiations exacerbated by the North–South divide (Roberts & Parks, 2007). This divide has evolved as world politics change, new economies emerge, and countries' interests and alliances shift (Caplet, Roberts, & Khan, 2015). AILAC's creation was partly in response to this changing landscape, in which several new groups emerged to express more independent positions within the G77 (Blaxekjær & Nielsen, 2014).

AILAC's origins date back to 2005 when the founding members and other countries began to collaborate on an *ad hoc* basis to discuss the climate negotiations and informal meetings. In 2006, Jose Alberto Garibaldi, an adviser to numerous Latin American governments on climate change, organized the 'Latin American and the Caribbean Workshops to Increase the Scale of Responses', in which several countries including Chile, Colombia, Costa Rica, Mexico, and Paraguay participated. These workshops were significant in providing an early space for these countries to begin to discuss, and attempt to coordinate, common positions. All of AILAC's founding members also participated in the Cartagena Dialogue for Progressive Action, which held its first official meeting in 2010, in Cartagena, Colombia. As an informal and private space, the Cartagena Dialogue allows participants to discuss the rationale behind each other's positions and explore areas of convergence and divergence (Araya, 2011; Cartagena Dialogue, 2010). It also includes members from Europe, Africa, Asia, and Small Island States. Chile, Colombia, Costa Rica, the Dominican Republic, Mexico, and Peru have participated actively. The space afforded by the Cartagena Dialogue allowed the future members of AILAC to develop important connections and build trust with negotiators from outside of the region and across the North–South divide. These interactions allowed these Latin American countries and other Dialogue participants to envision how to build bridges on difficult issues. Countries could advance their own agenda, while working simultaneously with nations from the Global North and South.

The complex negotiations and the limited size and capacity of the delegations of AILAC countries also led to increased coordination. Prior to its creation, delegates often acted as a single team, coordinating positions, sharing information, and occasionally representing each other in meetings (AILAC website, accessed September 2015). These experiences built trust, which strengthened the case to formalize the group. The origin of AILAC growing organically from a common narrative among negotiators rather than a top-down political agenda from politicians apparently represents a unique development at the UNFCCC.

Constructivists' explanations of how norms and values emerge and spread at a global level offer a useful way of explaining AILAC's development. Finnemore and Sikkink (2001) assert that interaction between people is shaped primarily by ideational factors, not only by material ones. Hurd (2008) states that constructivists focus on the social construction of actors, institutions, and events, which requires beginning with the assumption that how people and states think and behave is premised on their understanding of the world around them. Finnemore and Sikkink (2001) add that understanding how things are put together and how they occur is not just a descriptive exercise, but rather necessary since understanding the constitution of things will explain how they behave and what causes political outcomes. Wendt (1998) and Barnett and Finnemore (1999) describe constitution as causal, since how things are put together makes possible, or even likely, certain kinds of political behaviour and effects.

Constructivists distinguish between the 'logic of appropriateness', where actions are mediated by social norms and identities, and the 'logic of consequences', where actions are rationally chosen to maximize the interests of a state (Slaughter, 2011). These concepts are influential in organizational studies and characterize the logics of action of imperfectly rational actors and can help to understand and predict their behaviour (Schulz, 2014). March and Olsen (2009) state that the logic of appropriateness is a perspective that sees human action as driven by rules of appropriate or exemplary behaviour, organized into institutions. Rules are followed because they are seen as natural, expected, and legitimate. Actors seek to fulfil the obligations encapsulated in a role, an identity, a membership in a political group, and the practices and expectations of its institutions.

In AILAC's case, we can identify shifts between the logic of appropriateness and the logic of consequences at different stages of its development. Before AILAC's creation, the countries' actions were primarily a result of a convergence of norms and a common identity consistent with the logic of appropriateness, rather than an attempt to maximize their interests in keeping with the logic of consequences. At this point, AILAC countries had already established their identity as countries with an interest in challenging the North–South divide by offering new proposals and perspectives. This common identity was primarily based on material dimensions, given their middle-income country status, high level of vulnerability to climate impacts, and relatively low-carbon energy matrices due to their considerable installed hydroelectric capacity (Edwards & Roberts, 2015). Their identity also encompassed some symbolic elements, demonstrated by their increasing concern to become more relevant actors on the global stage.

Future AILAC members were among the first developing countries to make voluntary emission reduction pledges, despite having no obligations under the UNFCCC or the Kyoto Protocol to do so. In 2008, Peru was the first developing country to pledge to reduce to zero its net deforestation of primary forests by 2021.[7] This pledge illustrates how countries broke away from what used to be a G77 position red line, which was that commitments on emissions reductions were only applicable to developed countries (Caplet et al., 2015). They decided to act due to a growing sense of their own vulnerability to climate impacts and the hope to persuade larger countries, both developed and developing, to increase their own ambition (Garbiladi, 2014).

The creation of a group sought to overcome procedural limitations at the negotiations. During the negotiations, those delegates facilitating the proceedings regularly undertake individual consultations with groups to solve issues. As positions differed between subgroups of the G77, those who had no specific affiliation with a subgroup were only able to participate in consultations with the full G77 membership. Other countries which are part of subgroups, such as the African Group, the Least Developed Countries (LDCs) group, Alliance of Small Island States (AOSIS), and BASIC (Brazil, South Africa, China, and India), had the advantage of being consulted both at the general G77 meeting and at individual meetings where they were able to raise their concerns, including those where their position differed from the general G77 one. AILAC countries noted their positions were not being fully addressed under this arrangement.

The decision to formalize the group was intended to act as a counterweight to some G77 countries, including Brazil, and members of the Bolivarian Alliance for the Peoples of Our America (ALBA)[8] including Venezuela and Bolivia, which did not necessarily reflect the range of positions of other countries in Latin America (Edwards & Roberts, 2015). As these differences between AILAC countries and others in the G77 grew, so did the sense of exclusion when the consultation process did not include those not in a formal subgroup. Tired of insisting that they should also be consulted, these countries decided to formalize a new group. Paula Caballero and José Luis Balmaceda, at the time heads of delegation of Colombia and Chile, respectively, played a key role by bringing the group together and lobbying inside their ministries to secure approval. Colombia and Chile led the internal process between the foreign ministries of the founding AILAC members and the UNFCCC Secretariat to formally establish AILAC.

AILAC was launched at the UN climate conference in Doha, Qatar, in December 2012, allowing countries to develop a distinct identity and enhance their visibility within the G77 and with developed countries[9] (Roberts & Edwards, 2012). AILAC could now hold 'bilateral' meetings with other groups and countries, be consulted by those facilitating the negotiations, strengthen a common narrative, share technical and political expertise, and bolster their negotiating power. The decision to launch the group was in line with the 'logic of consequence', since the countries saw an opportunity to increase their visibility and promote their interests through their formal recognition.

In 2013, AILAC negotiators spent considerable time explaining their group to other countries. Negotiators faced strong pressures from other developing countries, given some of AILAC's positions are distinct from the rest of the G77. The biggest criticism from other developing countries came from the frequent misunderstanding of AILAC's position on the principle of Common but Differentiated Responsibilities and Respective Capabilities (CBDR + RC). Due to AILAC countries' readiness to put forward voluntary emission reduction pledges and their call for all countries to commit to their own emission reductions, many in the G77 interpreted AILAC's position as backing away from CBDR + RC. Some G77 countries understood this principle as implying that they did not have to put forward their own pledges. AILAC delegates spent considerable time explaining that their position did not refute CBDR + RC and the historical responsibility of developed countries to act first. AILAC's interpretation of CBDR + RC as being compatible with developing countries acting according to their own responsibilities and capabilities was often a point of contention.

This frequent questioning had the unintended consequence of accelerating the group's consolidation. AILAC countries also spent more time in internal meetings as every issue now required a joint position, which contributed to building up the group's capacity. The evolution of AILAC since 2012 can, therefore, be traced to actions on the logic of appropriateness, given the spread of norms and rules between its members. In the face of pressure from other countries and groups, AILAC strengthened its identity, sense of purpose, and belonging at the UNFCCC.

3. AILAC and the Paris Agreement

AILAC was able to make substantive contributions to the Paris Agreement and reacted positively to its adoption (La Vanguardia, 2015).[10] A number of the group's key positions, red lines, and 'bridging proposals' were considered in both the Agreement and accompanying decision 1/CP.21. During the 21st Conference of the Parties (COP 21) in Paris, AILAC members worked alongside other countries and groups. Costa Rica, Guatemala, and Honduras, for example, participated in the Climate Vulnerable

Forum (CVF). AILAC countries also participated in the High Ambition Coalition, an informal group that emerged at COP 21, which stems in part from the successful experience of the Cartagena Dialogue. Some of AILAC's positions overlapped with the CVF and the High Ambition Coalition, including the goal of limiting warming to 1.5°C (Edwards, 2016). AILAC was able to contribute to the process due its members' willingness to make commitments, and its capacity to hold constructive discussions with other countries. This section discusses five areas of the new agreement and how they correspond to AILAC positions, including differentiation, legal architecture and the INDCs, mitigation, adaptation, and finance.

3.1. Differentiation

The issue of differentiation between responsibilities of the developed and developing world has posed a significant challenge to the UNFCCC negotiations, as countries argue over who should reduce emissions first, how quickly, and by how much. The Paris Agreement represents an evolution in this area by building on the compromise at the COP 20 in Lima, Peru, in 2014, which adds the phrase 'in light of different national circumstances' after a reference to CBDR + RC. The Paris Agreement does not explicitly mention the listing of countries in Annexes I and II of the UNFCCC, but rather developed and developing countries. Here, differentiation is based largely on 'self- differentiation', wherein countries decide for themselves on which group they belong to (International Institute for Sustainable Development, 2015).

AILAC was largely in favour of the treatment of differentiation in the Paris Agreement (despite strong reservations about the legal architecture and the INDCs addressed in the next section). The group has stressed a preference for a flexible interpretation on CBDR + RC, emphasizing the importance of equity and encouraging *all* countries to take on commitments to reduce their emissions (*El País*, 2012).

By assuming responsibility for their emissions, AILAC countries differentiate themselves from other developing countries, while simultaneously demanding action by developed ones. AILAC's flexible interpretation of CBDR + RC lines up strongly with AOSIS, some LDCs, Mexico, and the EU. Together with some developing countries, including Small Island States, AILAC members have been willing to put forward unconditional emission reduction pledges, alongside further actions depending on international support. This approach rests on the idea that collective action reduces impacts and provides a collective benefit. This vision was more widely accepted in the Paris negotiations.

AILAC's framing of CBDR + RC has at times put it at odds with China, Venezuela, and Saudi Arabia, which focus on their 'right to develop' without constraints. Since more than half of global emissions now come from developing countries, AILAC stated that the principles of the Convention should be applied in a contemporary context and evolve with changing national circumstances (AILAC, March 2013). While the principles are enduring, they should allow for a broad range of differentiated commitments for all countries. For AILAC, a key part of the equity concept is about the right of vulnerable communities to live and prosper without climate impacts threatening their own survival and prosperity. The Paris Agreement text is consistent with AILAC's position that an agreement reflects respective capacities and responsibilities that are differentiated: developed countries still lead, but *all* countries act to the best of their ability with the necessary support from developed countries.

3.2. Legal Architecture and the INDCs

The Paris Agreement includes a legally binding requirement for Parties to prepare, communicate, and maintain successive Nationally Determined Contributions (NDCs) and pursue domestic mitigation measures. The Agreement also includes five-year cycles for increasing ambition, starting with a facilitative dialogue in 2018 and a global stock take in 2023 to monitor the progress towards the long-term goals.

AILAC was strongly in favour of a legally binding agreement for all Parties, a position it shared with other groups including the LDCs, AOSIS, and the EU. It argued that the Agreement must serve as a catalyst for national climate action, incentivizing countries' ambition, and providing a structure to support it. In a September 2014 submission, AILAC called for a legal anchor that would link national contributions to the legal agreement, turning them into legally binding obligations. It saw the need for all countries to include cycles of individual commitments for mitigation, adaptation, and finance to incentivize action with clear targets (AILAC, September 2014). AILAC said that a short-term commitment cycle is the most effective way to create an agreement that can trigger countries to frequently revisit and enhance their ambition (AILAC, August 2015). This position was shared with AOSIS, which has supported shorter and synchronized commitment cycles to avoid locking in low ambition for decades (AOSIS, 2015a). This was essentially the Paris outcome.

The September 2014 submission by AILAC outlined how to anchor the INDCs so each country is legally bound to implement its national contribution without needing to ratify the INDCs of other countries (AILAC, September 2014). AILAC wanted the agreement to include an explicit reference stating that all countries commit to implement what was set forth in their INDC, while being subject to the Convention's transparency requirements. AILAC did not originally support the INDC concept since it entailed too great a focus on a bottom-up system. Through its legal submission, AILAC sought to preserve as much as possible the top-down elements that could be reinserted into a new agreement. The group added that the proposed architecture would be underpinned by both the principles of the Convention and a general principle of non-backsliding, by which no country could withdraw its contribution or make a less ambitious contribution than its previous one. A principle of gradual scale-up would force nations to increase their commitments. This is largely the picture that unfolded at COP 21. The agreement includes the INDCs that each Party would commit to implement.

AILAC also called for a compliance mechanism not only to help developing country Parties meet their pledges, but also to include sanctions for recurring promise-breakers. It proposed that a compliance mechanism should be non-confrontational, transparent, and flexible (AILAC, September 2014). Although not all of AILAC's criteria for a compliance mechanism were included, the mechanism itself was established in the agreement.

3.3. Mitigation and long-term goals

The Paris Agreement aims to limit the increase in the global average temperature to well below 2°C above pre-industrial levels and to pursue efforts to limit the temperature increase to 1.5°C. The Agreement also includes a long-term mitigation goal to operationalize the global temperature goal, calling for net-zero emissions. In its Article 4, it states that Parties aim to 'achieve a balance between anthropogenic emissions by sources and removals by sinks of greenhouse gases in the second half of this century'. In October 2015, AILAC called for an agreement to 'establish a quantified, long-term trajectory, as recommended by the latest science ...' (AILAC, October 2015). AILAC's position on the need for mitigation goals to be driven by climate science is also shared by AOSIS, the LDCs, Mexico, and the EU, with some important differences regarding time frames. In AILAC's opening statement at COP 21, it called for the new agreement to include 'ambitious targets built around the 1.5 degree Celsius long term temperature goal' – an aspiration it shared with AOSIS (AILAC, November 2015; AOSIS, 2015b; see also Alfaro-Pelico, 2012).

3.4. Finance

The Paris Agreement includes a legal obligation for developed countries to provide climate finance, and states their leading role to mobilize it, highlighting the importance of public funds and the need to balance support for mitigation and adaptation. It also includes an obligation for developed countries to biennially provide forward-looking quantitative and qualitative information on climate finance. The Agreement does not refer to specific levels of finance, but rather decision 1/CP.21 on its adoption references the joint goal to mobilize a floor of US$100 billion annually by 2020, which is to be extended to 2025 when a new goal will be set. Developing countries had hoped that the floor of $100 billion would be surpassed immediately by 2020. The Agreement encourages developing country Parties to provide finance voluntarily. The need for public and grant-based resources for adaptation is also recognized, but with no clear quantified target, making such a provision appear ambiguous (Roberts &Weikmans, 2015).

According to Luis Fierro Carrion, former finance adviser in AILAC's support unit, the group concluded a successful negotiating strategy on finance at COP 21. Since 2014, AILAC has defined its following priorities on climate finance, including a collective quantified goal that takes as a 'floor' the existing commitment to mobilize US$100 billion per year from 2020, a qualitative long-term goal that would lead to all investments being gradually directed towards low-carbon and climate-resilient development, an obligation for developed countries to communicate *ex ante* quantified information on finance, and finally promoting more balanced finance for adaptation (approaching the levels provided for mitigation). These priorities were reflected in the Paris Agreement and decision 1/CP.21 on its adoption, but without binding language. All these priorities were shared by the rest of the G77 subgroups, including the LDCs and AOSIS, which is why AILAC orchestrated its strategy through the G77, by incorporating most of its priorities on finance in the submissions presented by the G77 (Fierro Carrion, 2016).

AILAC's positions on climate finance, like the rest of the G77, asserted that countries with less capacity to act must be provided with the necessary means of implementation, including the transfer of technology and capacity building, and that developed countries need to provide adequate and predictable levels of finance. However, an important difference made the G77's negotiating strategy less effective. AOSIS, the LDCs, and the African group strongly advocated for explicit references to their region or group in the agreement and decision 1/CP.21 to secure priority consideration for both access to and allocation of financial resources. AILAC joined by other Latin American countries, including Uruguay, opposed any special listing that would result in the exclusion of Latin America. This difference was overcome with a reference to the countries with the least capacity only in the paragraphs on the need to facilitate access to multilateral funds, and excluding any prioritization on allocation or scale of resources.

Since the Durban climate conference in 2011, AILAC members were among the main proponents of defining clear pathways for scaling up finance that ensure the fulfilment of the commitment undertaken by developed countries. Nonetheless, AILAC, Mexico, and the Dominican Republic stepped away from the narrative that developing countries could only be recipients of finance. In a 2012 submission, Colombia, Costa Rica, Dominican Republic, and Peru proposed a regime with universal contributions to the Green Climate Fund, according to countries' particular financial capacities and consistent with the principle of CBDR + RC (Colombia et al. submission, May 2012).

AILAC's September 2014 submission on the legal architecture stated that '[d]eveloped countries, which are historically responsible and highly capable, need to scale up the provision of resources, technology and capacity building for implementing adaptation action in particularly vulnerable developing countries', which includes AILAC members. For AILAC, 'the provision of resources, technology and capacity building must also to be tailored to the specific needs of middle income countries, which still harbor large vulnerable populations and have limited means to fund adaptation efforts with their own means' (AILAC, September 2014). This legal submission raised two of the most difficult issues that required intense negotiations within the G77 on finance: who contributes the finance and who receives it.

The issue of who contributes was one of the most controversial of the whole negotiations in Paris, and was included in the section on finance via the operationalization of the principle of CBDR + RC. Developed countries insisted on the inclusion of language that includes 'others in a position to do so' in the same sentence describing their obligation to provide resources. Meanwhile, most of the G77 opposed any reference to any kind of contributions – including voluntary – by non-Annex I countries. Despite divergent opinions on the issue of differentiation, AILAC's flexibility on the inclusion of references to voluntary contributions by 'any other countries with high capacity and in a position to do so' (AILAC, September 2014) opened up the space for reaching a final agreement on this matter. At the same time, the highly sensitive issue of vulnerability and prioritization of recipients, where AILAC was particularly active along with other Latin American countries including those of the ALBA group, required a significant amount of time and effort within the G77 and added an additional layer of complexity to the overall negotiation on the section on finance.

Although AILAC's positions on finance were largely reflected in the agreement, the lack of binding language and clarity on the roadmap as to how the US$100 billion per year would be reached cast a shadow over the advances achieved.

3.5. Adaptation

The Paris Agreement establishes a long-term goal to enhance adaptive capacity, strengthen resilience, and reduce vulnerability to climate change (Article 7). Under the Agreement, every country will engage in adaptation planning processes, while mandating a balance in finance between mitigation and adaptation, ensuring that funds for the latter will grow (Morgan, 2015).

AILAC was one of the main proponents of a global goal on adaptation and advocated for including adaptation as part of the national contributions through the creation of National Adaptation Plans (AILAC and Mexico, 2014). AILAC and Mexico's joint statement on adaptation outlined that a new agreement would need to include a global goal on adaptation, collective and individual adaptation commitments of nations, and means of implementation specifically addressing adaptation action in vulnerable countries. AILAC argued that a global goal on adaptation would provide a vision for the international community's efforts towards a climate-resilient planet, which could be assessed through a set of metrics and indicators to track, for example, the reduction of settlements in situations of risk, the reduction of economic losses, and the increase in national funding for adaptation (AILAC and Mexico, 2014).

While AILAC sometimes appears to hold positions different from the G77, it frequently overlaps with other groups within this bloc. AILAC's position was for some time similar to the Africa group and the Like-Minded Developing Countries group in wanting to include adaptation in the INDCs. AOSIS and the EU held a different position in which they opposed the inclusion, arguing that adaptation is essentially a local issue, and not a commitment of countries towards the international community. In the end, adaptation was included as a voluntary component of the INDCs, and later adaptation communications were included in the agreement. These communications to be submitted by Parties may 'include its priorities, implementation and support needs, plans and actions' on adaptation (Paris Agreement, Article 7.10).

Despite AILAC's positions on adaptation being largely met, the dearth of available funding for adaptation and limited transparency for existing sources reveal a difficult situation for how developing countries will implement the Paris Agreement's provisions on adaptation (AdaptationWatch, 2015).

4. Can AILAC continue to build a brighter future post-COP 21?

AILAC is now a significant player at the UNFCCC. Its decision to work across the North–South divide, which helped secure the Paris Agreement, has been vindicated, with largely positive outcomes on all five areas reviewed here. The group's future prospects appear positive, but substantial challenges remain.

In a key development, AILAC created an advisory group in 2014 to support its negotiators, which came in response to other negotiating groups able to call on their own support units.[11] The support unit provides policy and technical advice and offers logistics and communications support. AILAC has benefited from negotiators transferring to the support unit, which has proved a useful means to retain experience and provide consistency and continuity.[12]At the same time, the decision-making required to manage the support unit inspired the creation of a governance committee, composed of two delegates per country, which is the main administrative and political decision-making body for AILAC.

Backed by the support unit and a clearer organizational structure through its governance committee, AILAC was able to substantially increase its level of participation and visibility at the UNFCCC. In the run-up to the Paris conference, AILAC increased its contributions considerably by making a high number of submissions and statements, demonstrating its engagement and its proactive agenda.[13]

AILAC's increasing visibility and impact at the negotiations also reveal some tensions pointing to the difficulties of becoming a group. AILAC has been regarded by some other countries as a threat to their own positions, resulting in tougher negotiations and the need for greater effort to clarify and hone the group's positions. The spontaneous nature of AILAC countries' previously informal cooperation was no longer viable. The formalization of the group inevitably meant that AILAC had to follow certain internal norms and procedures, rendering the negotiations more cumbersome. While this has helped raise the profile and internal technical capacity of the group, and makes it structurally more similar to other negotiating groups such as AOSIS, it also means that joint positions no longer happen organically and require more coordination between group members. Being a group can also limit the flexibility with which members can create joint positions and submissions with non-AILAC members. Although there are exceptions (e.g. AILAC and Mexico's 2014 joint statement on adaptation), these undertakings are now slower.

In common with other groups, asymmetries between AILAC members – with some members playing a more assertive role than others – can lead to internal tensions in the group. There are also challenges to establishing well-defined common positions, which outsiders may view as a sign of disharmony or a potential weakness. For example, on the issues of how human rights would be included in the Paris Agreement or whether a 2°C or a 1.5°C temperature goal would be backed, there were some challenges in establishing common positions.

Following COP 21, it will be important to ensure that AILAC's relatively new norms and procedures compared to that of more established groups are as minimally restrictive as possible, so that joint positions with other countries can be generated. The relatively small size of AILAC country delegations and rapid turnover of some negotiators and ministers also present challenges in terms of ensuring that sufficient levels of experience and knowledge are retained and avoiding becoming overstretched during UNFCCC meetings.

AILAC does not currently hold press conferences during UNFCCC meetings, unlike many other groups and countries. Holding press conferences with journalists could help increase the group's visibility. Moreover, given that these press conferences are normally live-streamed on the UNFCCC website, it might help increase the level of awareness and visibility of the group with domestic media and even public audiences, which currently know little about their member's UNFCCC activities.

In 2015, AILAC was successful in attracting two new members to the group: Paraguay and Honduras. AILAC's ability to attract new members is a positive endorsement of its participation at the UNFCCC. Attracting additional members such as Argentina, Uruguay, or El Salvador would be a significant achievement and worth exploring. AILAC's influence could also benefit from enhancing its ties with other countries in the Cartagena Dialogue, High Ambition Coalition, and CVF.

One of the principal challenges facing the group is whether AILAC's pro-ambition rhetoric can stand up to the realities of national and international politics. Although most of the AILAC countries have put forward emission reduction pledges for pre-2020 and post-2020, and various national climate policies, the level of implementation remains uneven (Edwards & Roberts, 2015; Ryan, 2012). Furthermore, sustainable development policies across Latin America are often undermined by policies in other sectors, such as industry, mining, energy, or transport (Economic Commission for Latin America and the Caribbean, 2015). It is also unclear whether the region's economic and long-term climate and environmental goals are compatible. Many AILAC countries, including Chile, Colombia, and Peru, are also part of the Pacific Alliance trade group. The allure of commercial and investment opportunities with Asia and the tendency to rely on natural resources might undermine sustainable development objectives (Edwards & Roberts, 2015).

AILAC can be considered a willing victim of its own success. As the former head of its support unit, Andrés Pirazzoli, commented, AILAC has been pushing for a high ambition global regime that can help trigger domestic action on climate change (personal communication, June 2015). Therefore, scrutiny of AILAC's rhetoric at the negotiations by other countries and non-state actors, and whether that rhetoric matches domestic action, may appear to be one way to spur ambition and action at the national level. This is especially relevant given that the Paris Agreement lacks a compliance mechanism with teeth. The international community and non-state actors – in particular, national and transnational civil society groups – will need to play an increasingly important role to ensure countries match their rhetoric with action and fulfil the Paris Agreement's goals (Falkner, 2016).

All AILAC members have now submitted an INDC or NDC. The level of ambition of the mitigation pledges in the contributions varies considerably. For example, Chile made an unconditional target of a 30% reduction in CO_2 emissions-intensity of the gross domestic product below 2007 levels by 2030, and of 45% with further international support. Among the most ambitious pledges, Costa Rica made an unconditional target to keep net emissions below 9.37 $MtCO_2e$ by 2030.[14] Analysis by the independent group Climate Action Tracker (October2015) rated Chile's submission as 'inadequate', as it is inconsistent with limiting global warming to below 2°C. Costa Rica's INDC, however, is rated as 'sufficient' and one of the best in the world evaluated by the group, although this rating could be improved if Costa Rica renewed its commitment to have zero carbon emissions by 2021. At present, however, there appears to be little willingness to revise and resubmit more ambitious INDCs or NDCs.[15] In the case of Chile, there is less enthusiasm to revise its contribution, given the forthcoming elections in 2017. In Colombia, the peace process has drained away attention from other policy efforts like climate change. The ability of AILAC members to implement the current targets and policies in their INDCs/NDCs and increase the level of ambition will be contingent on whether there is sufficient national backing for such an endeavour, and crucially whether the support from developed countries especially on finance, technology transfer, and capacity building is forthcoming.

AILAC has gained traction at the UNFCCC due to its significant contributions to the Paris Agreement. Given how quickly the group came together and its small size, AILAC's impact on the Agreement is remarkable. Two of the biggest challenges for the group's legitimacy going forward will be, first, how quickly the other members can ratify the Paris Agreement, and second, the extent to which AILAC countries actually implement their pre-2020 voluntary emission reduction pledges and longer term INDCs/NDCs, which is uncertain. Implementation, of course, is also contingent on developed countries fulfilling their commitments on finance, itself somewhat unpredictable. Domestic political and economic turmoil threaten the focus and resolve of this emerging group of developing nations to climate leadership. AILAC may have established itself as a bridge builder at the negotiations; now it has to demonstrate it can build durable bridges towards a low emission and resilient future.

Acknowledgements

We thank Paula Caballero, Monica Araya, Andres Pirazzoli, Alexa Kleysteuber, Irene Suarez, Alejandro Rivera, Gilberto Arias, and Jose Alberto Garibaldi for discussing AILAC with us. All errors remain our own. The research was primarily conducted at Brown University.

Disclosure statement

No potential conflict of interest was reported by the authors.

Notes

1. Some AILAC members are also part of the Central American Integration System (SICA) and the Coalition for Rainforest Nations.
2. World Resources Institute CAIT Equity Explorer (CAIT.WRI.org, 2012 data accessed 22 August 2016). When land-use change is included, all AILAC members were below the global average except Paraguay.
3. Authors' personal communication with Sarah Thacker, formerly of the UK's Department of Energy and Climate Change, February 2013
4. Authors' personal communication with John Ashton, former Special Representative for Climate Change at the UK Foreign and Commonwealth Office, January2013
5. Authors' personal communication with the author Mark Lynas, December 2011
6. Paraguay is a member of Mercosur, which is considered less open to free trade than the Pacific Alliance.
7. Chile, Colombia, and Costa Rica also made voluntary pledges. Panama, Paraguay, and Guatemala did not formally submit pledges to the Copenhagen Accord.
8. Alianza Bolivariana para los Pueblos de Nuestra América.
9. Authors' personal communication with Paula Caballero, former Director for Economic, Social, and Environmental Affairs in the Ministry of Foreign Affairs of Colombia, September 2013.
10. At the time of writing in September 2016, Peru and Panama have ratified the Paris Agreement and some other AILAC members also plan to ratify in 2016.
11. Other groups including AOSIS, LDC, and the Like-Minded Group of Developing Countries have technical support units.
12. Authors' personal communication with Alexa Kleysteuber, former member of the AILAC Support Unit, June 2015.
13. All submissions are available on the UNFCCC's website http://unfccc.int/documentation/submissions_from_parties/items/5900.php.
14. See UNFCCC INDC database http://www4.unfccc.int/submissions/INDC/Submission%20Pages/submissions.aspx.
15. When countries ratify the Paris Agreement, they are required to submit a confirmed pledge, which is then known as an NDC (no longer just 'intended').

References

AdaptationWatch. (2015). *Towards mutual accountability: The 2015 adaptation finance transparency gap report.* Retrieved from http://adaptationwatch.org

AILAC. (2013, March 1). *Submission on the ADP – Planning of Work in 2013.* Author. Retrieved from http://ailac.org/en/remisiones-documentales/

AILAC. (2014, September). *Submission on the legal architecture and structure of the elements of the 2015 Agreement.* Author. Retrieved from http://ailac.org/en/remisiones-documentales/

AILAC. (2015, August 31). *Opening statement for the ADP 2 10.* Bonn, Germany: Author. Retrieved from http://ailac.org/en/remisiones-documentales/

AILAC. (2015, October 19). *Opening statement for the ADP 2 -11.* Bonn, Germany: Author. Retrieved from http://ailac.org/en/remisiones-documentales/

AILAC. (2015, November 29). *Opening statement for the ADP 2 12*. Paris, France: Author. Retrieved from http://ailac.org/en/remisiones-documentales/

AILAC and Mexico. (2014, October 18). *Adaptation in the ADP*. Joint submission of AILAC and Mexico: Author. Retrieved from http://ailac.org/en/remisiones-documentales/

Alfaro-Pelico, R. I. (2012). *Small island developing states and climate change: Effects, responses and positions beyond Durban* (Working Paper). Madrid: Elcano Royal Institute.

AOSIS. (2015a, October 19). Small Islands: Co-Chairs' text falls short in Bonn. *Press Release*. Retrieved August 16, 2016, from http://aosis.org/small-islands-co-chairs-text-fall-short-in-bonn/

AOSIS. (2015b, November 29). Paris must show global solidarity to tackle climate change. *Press Release*. Retrieved August 16, 2016, from http://aosis.org/paris-must-show-global-solidarity-to-tackle-climate-change/

Araya, M. (2011, February 2). *The Cartagena Dialogue: A Sui Generis Alliance in the Climate Negotiations*. Intercambio Climático. Retrieved from http://intercambioclimatico.com/en/home/item/286-85the-cartagena-dialogue-a-sui-generis-alliance-in-the-climate-negotiations.html

Barnett, M. N., & Finnemore, M. (1999). The politics, power, and pathologies of international organizations. *International Organization, 53*, 699–732.

Blaxekjær, L. Ø., & Nielsen, T. D. (2014). Mapping the narrative positions of new political groups under the UNFCCC. *Climate Policy, 15*(6), 751–766.

Caplet, D., Roberts, J. T., & Khan, M. R. (2015). *Power in a warming world. The New global politics of climate change and the remaking of environmental inequality*. Cambridge: MIT Press.

Cartagena Dialogue. (2010). *Chairman's statement*. Third meeting of the Cartagena Dialogue for progressive action, San Jose, Costa Rica, October 31–November 2. Retrieved from http://switchboard.nrdc.org/blogs/sclefkowitz/Cartagena/Chairman's_Statement_COSTA_RICA%20Nov%202010.pdf

Climate Action Tracker. (2015, October 29). *Large South American governments could take more climate action: Analysis*. Retrieved January 20, 2016, from http://climateactiontracker.org/news/234/Large-South-American-governments-could-take-more-climate-action-analysis.html

Economic Commission for Latin America and the Caribbean. (2015). *Latin America and the Caribbean: Looking ahead after the millennium development goals*. Regional monitoring report on the millennium development goals in Latin America and the Caribbean, Santiago de Chile.

Edwards, G. (2016, February 4). *Can Latin American diplomacy at COP21 spur interest in the Paris deal back home?*Nivela Retrieved from http://www.nivela.org/articles/can-latin-american-diplomacy-at-cop21-spur-interest-in-the-paris-deal-back-home/en

Edwards, G., & Roberts, J. T. (2015). *A fragmented continent: Latin America and the global politics of climate change*. Boston: MIT Press.

Falkner, R. (2016). The Paris Agreement and the new logic of international climate politics. *International Affairs, 92*(5), 1107–1125.

Fierro Carrion, L. (2016, February 3). *AILAC achieved success in its climate finance priorities in the Paris Agreement*. Linkedin Blog Post. Retrieved September 10, 2016, from https://www.linkedin.com/pulse/ailac-achieved-success-its-climate-finance-priorities-luis-fierro?trk=prof-post

Finnemore, M., & Sikkink, K. (2001). TAKING STOCK: The constructivist research program in international relations and comparative politics. *Annual Reveiw Political Science, 4*, 391–416.

Friedman, L. (2013, January 22). *New alliance attempts to bridge North-South gap in U.N. climate talks*. ClimateWire. Retrieved September 10, 2016, from http://ccap.org/assets/New-Alliance-Attempts-to-Bridge-North-South-Gap-in-UN-Cliamte-Talks_EE-ClimateWire-Jan-2013.pdf

Garbiladi, J. A. (2014). The economics of boldness: Equity, action, and hope. *Climate Policy, 14*(1), 82–101.

Hurd, I. (2008). Constructivism. In C. Reus-Smit & D. Snidal (Eds.), *The Oxford handbook of international relations* (pp. 298–316). Oxford: OUP.

International Institute for Sustainable Development. (2015, December 15). *Summary of the Paris climate change conference 29 November–13 December*. Vol. 12 No. 663. Retrieved March 5, 2016, from http://www.iisd.ca/vol12/enb12663e.html

La Vanguardia. (2015, December 12). *Países AILAC piden 'seriedad' en la aplicacióndelacuerdoclimático de París*. Retrieved http://www.lavanguardia.com/politica/20151212/30764368697/paises-ailac-piden-seriedad-en-la-aplicacion-del-acuerdo-climatico-de-paris.html

March, J. G., & Olsen, J. P. (2009, April). *The logic of appropriateness* (ARENA Working Papers). Retrieved August 3, 2016, from http://www.sv.uio.no/arena/english/research/publications/arena-working-papers/2001-2010/2004/wp04_9.pdf

Méndez, R. (2012, December 5). La tercera vía latino americana en la negociación del clima. *El Pais*. Retrieved from http://sociedad.elpais.com/sociedad/2012/12/05/actualidad/1354699047_259945.html

Morgan, J. (2015, December 14). *4 Signs the Paris Agreement is the start of a new era in international climate action*. World Resources Institute. Retrieved from http://www.wri.org/blog/2015/12/4-signs-paris-agreement-start-new-era-international-climate-action

Roberts, J. T., & Edwards, G. (2012). *A new Latin American climate negotiating group: The greenest shoots in the Doha Desert*. The Brookings Institution. Retrieved from https://www.brookings.edu/blog/up-front/2012/12/12/a-new-latin-american-climate-negotiating-group-the-greenest-shoots-in-the-doha-desert/

Roberts, J. T., & Parks, B. (2007). *A climate of injustice*. Boston: MIT Press.

Roberts, J. T., & Weikmans, R. (2015, December 22). *The unfinished agenda of the Paris climate talks: Finance to the global south*. Planet Policy Blog, Brookings Institution. Retrieved from http://www.brookings.edu/blogs/planetpolicy/posts/2015/12/22-cop21-unfinished-agenda-finance-global-south-roberts-weikmans

Ryan, D. (2012). *Informe sobre el Estado y Calidad de las Politicas Publicas sobre Cambio Climatico y Desarrollo en America Latina: Sector Agropecuario y Forestal. Plataforma Climatica Latinoamericana*. Retrieved from http://intercambioclimatico.com/es/publicaciones/17-informe-sobre-el-estado-y-calidad-de-las-politicas-publicas-sobre-cambio-climatico-y-desarrollo-en-america-latina-sector-agropecuario-y-forestal-2012/file.html

Schulz, M. (2014). Logic of consequences and logic of appropriateness. In M. Augier & D. Teece (Eds.), *Palgrave encyclopedia of strategic management* (p. 2). Basingstoke: Palgrave Macmillan.

Slaughter, A.-M. (2011). International relations, principal theories. In R. Wolfrum (Ed.), *Max Planck encyclopedia of public international Law* (p. 4). Oxford: Oxford University Press.

Submission by Colombia, Costa Rica, Dominican Republic and Peru. (2012, May 15). *Views on the work plan of the Ad Hoc Working Group on the Durban Platform for Enhanced Action*. Bonn, Germany. Retrieved from http://www4.unfccc.int/Submissions/SitePages/sessions.aspx

Wendt, A. (1998). Constitution and causation in international relations. *Review of International Studies, 24*(5), 101–118.

US-proofing the Paris Climate Agreement

LUKE KEMP

What can be done if a future US President withdraws from the Paris Climate Agreement? This is a critical question for an agreement which relies on universal participation for legitimacy. The US can and likely will ratify the Paris Agreement via a presidential-executive agreement. However, this could be easily reversed by a future President or an antagonistic Congress. A non-party US operating outside of the Paris climate architecture is a real threat, especially given Republican opposition to international climate policy. This article explores both what tools are available for addressing a non-party US in any climate agreement and what the Paris Agreement in particular can do. The Paris Agreement is extremely vulnerable to the withdrawal of the US, or any other major party. It possesses no non-party measures, although it can be amended to help avoid this Achilles heel. Through amendments to Article 6 a market link between subnational states in a renegade US and international carbon markets could be created. Ideally, a more semi-global approach with punitive carbon border tax adjustments could be taken to help US-proof an alternative climate agreement. The Paris Agreement will be fatally susceptible to a US dropout unless amendments to the treaty are made. Relying on the good will of a single president is short-sighted. Longer-term climate governance needs to take seriously the threat of non-parties, particularly if they are superpowers.

Policy relevance

This article explores how to craft an effective climate treaty without US ratification, including through reforms to the Paris Agreement. While the US is likely to be able to be able legally adopt the Paris Agreement through a presidential-executive agreement there is a distinct possibility of the US withdrawing and repealing such an agreement due to a change in president. This article suggests that the Paris Agreement is vulnerable to such an occurrence and thus the insights of this research are both timely and relevant to delegations and others involved with the UNFCCC. It will be of use to domestic US actors, as well as international negotiators and scholars in dealing with a non-party US and non-parties in general.

1. Introduction

If any future climate treaty is to be legally binding, it must be designed to deal with the very real possibility of US non-ratification. A key question for the 2015 Paris Climate Agreement is whether it is equipped to deal with the US not legally adopting it, or withdrawing under a future president. That is, can the Paris Agreement handle a 'non-party' US? The participation and/or leadership of the US is seen by many as being crucial, if not necessary, for the success of any climate agreement (Grundig & Ward, 2015; Purvis & Stevenson, 2010). But the US is caught in a 'ratification straitjacket': international legal engagement on environmental issues is significantly constrained due to a combination of unique domestic institutions and the problematic federal politics (Kemp, 2015a). The recently established 2015 Paris Climate Agreement under the United Nations Framework Convention on Climate Change (UNFCCC) has been designed to allow for the US to legally participate through a presidential-executive agreement. This would bypass congressional approval and not require ratification from the US Senate. However (as further explored in Section 2), there are numerous reasons to be prepared for the possibility of the US not legally adopting the Paris Climate Agreement or withdrawing under a future President. This is particularly true if a Republican rises to power in the coming decade. This would undermine the key achievement of the agreement: legitimacy through broad, if not universal, participation. It is therefore necessary for the world to consider how vulnerable the Paris Climate Agreement is to a non-party US and what reforms can be taken to address this potential weakness.

Despite the proliferation of literature on architectures for a 2015 agreement, the existing literature has not addressed the conundrum of the US ratification straitjacket and has rarely examined the management of non-parties in general. I (Kemp, 2015a) have examined how a climate agreement could remain effective and allow for US legal participation. But no-one has grappled with the other side of the equation: how a climate treaty could effectively operate without US legal participation. Given the existing political realities and Paris Agreement this is a significant gap with clear policy implications. I will address this void in existing literatures through pursuing three related research aims:

1. To provide an overview of the existing options that would allow for an effective climate treaty without US legal participation, or effective US participation in the absence of ratification.
2. To assess whether the existing Paris Climate Agreement can effectively manage a non-party US.
3. To provide recommended changes and alternatives to the Paris Climate Agreement that would better address a non-party US.

At this point, it is worthwhile to clarify a number of key terms. For the purposes of this article, an 'effective climate treaty' is any agreement which is likely to maintain sufficient ambition and give a reasonable chance of meeting the goals of the Paris Agreement (as enshrined in Article 2). 'Effective US participation' refers to the political and financial (not legal participation under international law) involvement of US actors within an international treaty that will lead to increased climate action and ambition within the US. Political feasibility in this case refers to institutional possibilities and the likelihood of acceptance by major powers.

I will proceed by first looking at the ability of the US to ratify the Paris Agreement and the likelihood that they will become a non-party. I will then provide a brief overview of climate policy in the US before examining how a climate agreement could cope with a non-party US, whether the Paris Agreement is US-proof, and finally how a climate agreement, including the Paris Agreement, could be designed to handle a non-party US.

2. Will the US become a non-party to the Paris Agreement?

The US has struggled to ratify international treaties, particularly environmental ones due to institutional and political barriers (Bodansky, 2015; Falkner, 2005; Kemp, 2015a; Schreurs, Selin, & VanDeveer, 2009). Article 2, section 2, clause 2 of the US constitution states that the president shall 'have the power, by and with the advice and consent of the Senate to make Treaties, provided two thirds of the Senators present concur'. Acquiring this supermajority vote in the Senate has proved incredibly difficult. A deep-seated scepticism of UN multilateralism, particularly on environmental matters (Brunnee, 2008; Patrick, 2002) along with lobbying and partisan politics, has made widespread political agreement on climate policy almost impossible in the short term (Bang, Hovi, & Sprinz, 2012; Purvis & Stevenson, 2010). Interestingly many treaties tend to pass through the Senate with few hurdles. The problem with others, particularly environmental treaties, is that they centre on more politically controversial topics. The ratification structure is not automatically a problem: it is only a major impediment when political opposition uses the Senate processes to delay, stall, and stop ratification. This 'ratification straitjacket' (Kemp, 2015a) was a sincere concern prior to the Paris Agreement and shaped the end outcome.

Presidential-executive agreements are an alternative method of international legal engagement that have the same legal force as Senate treaty ratification once enacted (Bodansky, 2015; Peake, Krutz, & Hughes, 2012). In contrast to Senate ratification a presidential-executive agreement only needs the consent of the Executive (Krutz & Peake, 2009). Presidential-executive agreements have been enacted to numerous multilateral environmental agreements (MEAs) including several protocols to the 1979 Long-Range Transboundary Air Pollution Convention (LRTAP) and the more recent 2013 Minamata Convention on Mercury. There is a general scholarly consensus that the US could engage with an international climate treaty through the use of presidential-executive agreements as long as it lies within the scope of the President's foreign affairs mandate and existing legislation (Bodansky, 2015; Chang, 2010). This means that the content of the treaty could not include legally binding emissions reductions targets or financial commitments.

The US is capable of legally adopting the Paris Agreement through a presidential-executive agreement. The Paris Agreement does not include binding commitments on financing, while mitigation commitments are nationally determined and non-binding. Parties simply have a legal commitment to 'prepare, communicate and maintain successive nationally determined contributions that it intends to achieve'. The content (or lack thereof) of the Paris Agreement means that it can be adopted through a presidential-executive agreement by Obama and would likely withstand a judicial challenge. Indeed, the Paris Agreement was designed to allow for exactly this. This fits my prediction (Kemp, 2015a) that Paris would be constructed as a pledge and review agreement with varied legality in order to allow for a presidential-executive agreement.

While promising, presidential-executive agreements should not be considered as a panacea to the problem of US participation. Presidential-executive agreements could be abolished by a future President through an executive order or by Congress via a contradictory later-in-time statute. While the same can also be done for binding commitments, such international legal obligations generally involve a domestic ratification and legislative process which requires broad political support. A presidential-executive agreement would be done to bypass domestic political consensus. Any future Republican president is likely to repeal such an agreement, effectively making the US a non-party. Indeed, many Republicans have already signalled their opposition to a presidential-executive agreement to Paris. A 2015 proposed Senate Resolution by Republican representatives Mike Lee and Mike Kelly calls for President Obama to submit the Paris Agreement for the advice and consent of the Senate as an Article II treaty.[1] EU diplomats have already expressed concerns that a future Republican will pull the US out of the Paris Agreement (Neslen, 2016).

US legal participation through executive agreements is contingent upon strong political will by the President. The Obama administration has enacted a number of promising domestic measures (which will be further discussed in Section 3.2) and has increased its rhetoric around addressing climate change. But it is questionable as to whether this will translate into the drastic act of enacting a sole-executive agreement on the international stage which could cause a political firestorm at home. The Clinton administration was similarly active prior to the Kyoto Protocol which it helped shape and design. However, the administration ultimately did not follow through with ratification. The threat of numerous reprisals by the Republican Party, including the withholding of funds for the Green Climate Fund (GCF) or domestic climate action programmes, could deter Obama from pursuing an executive agreement. Given this it is necessary to consider the real possibility and threat of the US not legally adopting, or adopting and later withdrawing from, the Paris Agreement. While domestic politics is the primary barrier here, domestic policy circumstances provide some key insights into how an international regime could effectively handle a non-party US.

3. The current state of US climate policy

3.1. The failure of federal climate policy

The most marked aspects of US climate policy are the lack of involvement in international legal instruments and absence of any central federal policy on mitigation. There has been a multitude of proposals for national emissions reductions schemes in the Senate and Congress, most of which have revolved around a cap and trade type architecture. This includes the McCain and Lieberman proposal at the 109th Senate session, and the Waxman Markey bill of 2010. Efforts have intensified as of late with six different carbon pricing proposals being put forward in the 113th Congress (2013–2014), five of which seek to establish a carbon tax (Ye, 2014). None of these proposals have been successful. It is indicative of both US climate politics, and the difficulty of crafting legislation in the US, that despite the sheer number of proposed bills the US still lacks a legislated national mitigation scheme. However, it should come as no surprise given that the US system is prone to gridlock (Bang, Froyn, Hovi, & Menz, 2007; Bang et al., 2012; Tjernshaugen, 2005) even when bipartisanship, including on issues such as energy security, exists (Bang, 2010).

In lieu of effective federal-level policy, the current executive has taken an increasingly proactive role. The Obama administration has made a number of promising advances forward such as the $3 billion US dollar pledge to the GCF, a bilateral climate arrangement with China (including cooperation on the phase-out of hydrochlorofluorocarbons, a potent greenhouse gas), and an intended nationally determined contribution (INDC) of reducing greenhouse gas emissions by 26–28% on 2005 levels by 2025.[2] In addition to these international actions, the Obama administration has also made extensive use of the Environmental Protection Agency (EPA) to enact a number of climate-related initiatives. These range from vehicle emissions standards through to the more recent Clean Energy Power Plan. The increasing activity of the executive on climate change has been matched by a multitude of ambitious initiatives by subnational US actors.

3.2. Subnational climate action in the US

In the absence of congressional climate action, there has been a blossoming of policies by states, regions, municipalities, and cities within the US. This includes such notable initiatives as the emissions trading system of the state of California (Schreurs, 2008). Many of the actions and goals of subnational

Table 1. A comparison of ambitious US city and state targets and the US INDC

Actor	Jurisdiction	Interim emissions reductions target(s)[a]	Long-term emissions reductions target
US	Nation state	26–28% below 2005 levels by 2025 (equivalent to – 14% to 19% on 1990 levels)	83% below 2005 by 2050 (equivalent to approximately 80% below 1990 levels)
Austin	City	Make all City of Austin facilities, fleets, and operations carbon neutral by 2020	Carbon neutral by 2050
California	State	Reduce to 1990 levels by 2020. 40% below 1990 levels by 2030	80% below 1990 levels by 2050
New York City	City	N/A	80% on 2005 levels by 2050
New York State	State	Reduce GHG emissions from the energy sector by 40% by 2030 based on 1990 levels	80% on 1990 levels by 2050
Seattle	City	Previously 7% below 1990 levels by 2012	Carbon neutral by 2050

[a]Emission targets are in GHG emissions unless stated otherwise. All targets are based on data gathered from the Carbonn database, the NAZCA database and cross-referenced against existing official documentation (where available).

jurisdictions are significantly more ambitious than national-level policies and targets. Table 1 provides a comparison of some of the most ambitious targets of US subnational jurisdictions against the existing US INDC. Importantly, US state policies in the past have served as forerunners of and models for later federal legislation (Rabe, 2008; Rabe, Roman, & Dobelis, 2005; Selin & VanDeveer, 2011) by acting as both laboratories for innovation (Burtraw & Woerman, 2012, 2013) and political advocates for national action. An example of this is the vehicle emissions standards enacted by California which were later adopted by the EPA and Obama administration as federal-level policy in 2010. Subnational action on climate change often outpaces federal-level policy in laggard countries, and the US is no exception to this trend (Lachapelle & Paterson, 2013).

The rise of subnational and executive-led actions is a clear reaction to federal gridlock. It is an alternative policy pathway to bypass the divided Congress and Senate (Farber, 2014). Indeed, Rabe has characterized this blend of subnational and federal initiatives as 'contested federalism' (2011). This analysis is supported by the design of existing federal climate policy. The Clean Power Plan sets different emissions rate goals for different states and provides the states with considerable flexibility in how they choose to meet their targets. Similarly, the regulation of carbon dioxide under the EPA Clean Air Act relies heavily on state interpretation and implementation (Rabe, 2011, p. 505). Executive policy on climate change has come to rely on US states for inspiration and implementation in the age of contested federalism.

3.3. The lessons of 'contested federalism'

Current US climate policy largely consists of fragmented subnational actions accompanied by a number of executive measures, all done in the shadow of Congressional gridlock. This has important limitations. Executive actions taken by the White House rely upon the President's mandate and existing legislation. In contrast, federal action can create entirely new legislation, such as a binding cap on

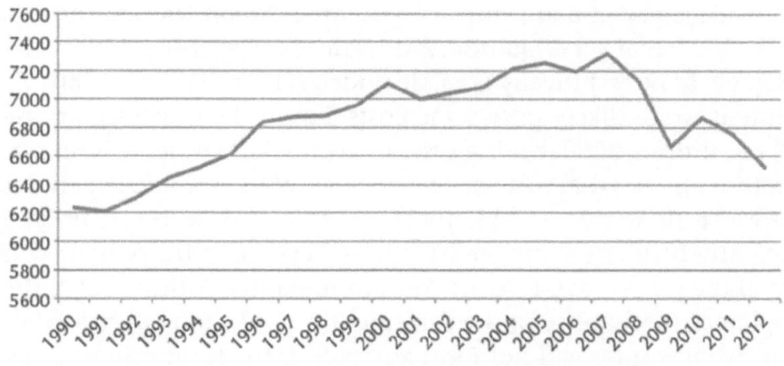

Figure 1. Trends in US GHG emissions, 1990–2012 (EPA, 2014)

emissions. Despite existing actions, the overall national emissions of the US have not experienced any substantial decline, as shown in Figure 1.

Contested federalism also provides unique opportunities. While ratification may be highly unlikely, linkages could still be established between subnational actions and a wider international architecture. The key in addressing a non-party US is to utilize tools that allow for the involvement, recognition, and financial links to progressive US actors.

4. Tools for managing a non-party US

This section provides an overview of the different options that can be used to help address US non-party status under a climate agreement. Section 5 then explicitly examines these tools in relation to the Paris Agreement to determine how robust or vulnerable the agreement is to a non-party US. The mechanisms discussed in this section are intended to coerce or incentivize eventual US legal reengagement and/or encourage mitigation activities within the US.

4.1. Non-party measures

Currently the Paris Climate Agreement has no non-party measures. There are no direct ways to handle a non-ratifying party or one that has withdrawn without amending the agreement. If such amendments are to occur, it is useful to understand what options are available.

4.1.1. Penalties

Punitive provisions for non-parties can take numerous forms but share the common aim of penalizing non-parties and therefore preventing free-riding. They usually take the form of trade restrictions that levy imports on the basis of involving either carbon border tax adjustments (BTAs) or the surrendering of emissions permits (Aakre, 2014). There is clear successful legal precedent for such non-party measures and their application in other MEAs. Trade in ozone-depleting substances with non-parties was banned under paragraph 4, Article 4 of the Montreal Protocol. Such measures have legal precedent in numerous other MEAs. BTAs have been repeatedly discussed within the EU and US (Aakre, 2014) and could potentially provide a way of managing any potential carbon leakage and addressing free rider concerns (Barker et al., 2007; Branger & Quirion, 2014; Purvis & Stevenson, 2010). The implementation of such measures is likely to be technically possible under WTO rules (Veel, 2009; Zhang, 2009) and a credible threat. While it would likely pose some costs due to the difficulty in assessing the carbon content of imports, monitoring and labelling systems are feasible (Ismer & Neuhoff, 2007; Pauwelyn, 2013). Existing literature suggests that cost and effectiveness of BTAs would largely depend on their particular design (van Asselt & Brewer, 2010). Whether parties would be willing to utilize BTAs, particularly against the US, is an entirely different matter. BTAs have never been implemented and have fallen off the political agenda of the EU (Kulovesi, 2014). Most importantly, Article 15.2 of the Paris Agreement states that the compliance mechanism of the agreement is to be 'non-adversarial and non-punitive'. BTAs and other punitive measures will not form any part of the future compliance mechanism and are highly unlikely to be part of any future non-party provisions. Punitive non-party measures, while theoretically useful, are currently not a feasible tool.

4.1.2. Facilitative measures

Non-party measures need not be punitive by nature and instead can be much more facilitative and inclusive of non-parties. Haites, Yamin, and Höhne (2013) suggest a provision for non-party recognition where a country with equivalent measurement, reporting and verification (MRV) and financial commitments can make a request to the COP to be recognized under the treaty. If accepted they would be expected to meet their proposed targets and gain access to flexibility mechanisms and carbon markets under the agreement. The development and use of carbon markets would appear to be a key incentive for US participation given their insistence upon their inclusion within the Kyoto Protocol and continued advocacy for market-based mechanisms. However, such facilitative measures rely on the consent of the president. If a future president is the reason for withdrawal from the agreement, then they are unlikely to want to engage through such facilitative provisions. Accordingly, facilitative measures are unlikely to be useful in managing a non-party US.[3]

Different non-party provisions, both punitive and facilitative, could be combined for flexibility and maximum effectiveness. For example, mixing border tax adjustments with non-party recognition and incentives would encourage non-ratifying parties to adopt comparable measures and request recognition so as to avoid trade restrictions and make the most of potential commercial opportunities. A carrot and a stick together will be more effective than either alone. One example of this in practice would be the 2013 *Minamata Convention on Mercury* which under Article 6(b) prohibits the trade of mercury with non-parties unless they provide certification proving that they meet the provisions on Articles 10 and 11 on sound storage and the protection of human health and the environment. A similar arrangement with the measures outlined above could both allay competitiveness concerns for Parties under a treaty without the US and still allow for the US to effectively engage despite a non-party status.

4.2. Facilitating bottom-up US action

A multilateral agreement is inherently a legal contract between nation states. This excludes the direct legal involvement of subnational actors such as states, cities, and local governments. However, there are other ways of incorporating subnational actors into an agreement. These range from the recognition of actions and mitigation measures by subnational and non-state actors through to direct financial connections.

4.2.1. Subnational facilitation

One way to involve subnational actors is to employ an international framework for recognition and facilitation. Linking the multilateral UNFCCC regime to subnational and non-state actions could help all three to realize their potential (Betsill et al., 2015). Since this requires little to no legal or financial engagement or resource expenditure it is the easiest and most politically feasible way of engaging subnational actors within a non-ratifying US. Technically, registries already exist that officially and publicly record the actions of subnational and non-state actors. These include the *carbonn* climate registry and the Non-State Actor Zone for Climate Action (NAZCA), both of which have recorded the initiatives of US cities, states, and companies. However, simply recording actions is not enough to encourage policy diffusion and upscaling of best practice. Chan and Pauw (2014) advocate for a comprehensive 'global framework for climate action' that will have numerous functions including the mobilization of new or improved initiatives, recording initiatives into a publicly available register, and conducting a periodic review of the effectiveness of different initiatives. The aim of such a framework would be not only to recognize subnational initiatives, but also steer and orchestrate them towards greater ambition and effective implementation.

Yet, there is no guarantee that such a platform would be effective. It could be argued that there is already sufficient knowledge of subnational actions within the US, particularly through international networks such as Local Governments for Sustainability (ICLEI) and the C40. The use of benchmarking and periodic review, while useful, does not always ensure compliance or diffusion. Going beyond facilitation towards more tangible financial links is one way to more strongly embed and aid subnational US action in an international agreement.

4.2.2. The subnational market link

While states in the US and subnational actors cannot directly ratify a multilateral agreement, they can still be involved by way of markets and financial transactions. A 'subnational link' could include both a project-based financing flexibility mechanism targeted towards subnational actors and access to international carbon markets for subnational actors. The former could be an expanded and improved CDM geared towards aiding subnational entities in meeting their commitments (LefÈvre, 2012). Directing financing channels and the leveraging of private finance through this mechanism could further accelerate subnational mitigation (Lefèvre & Wemaere, 2009). The latter measure could simply be an option to allow for subnational actors in non-parties to access international carbon markets by connecting their own carbon pricing schemes and directly trading certified permits on the international market. Unlike in current circumstances for parties using the Kyoto ETS, the permits would accrue to the state or implementing subnational actor, instead of the national government. This system could be used for parties as well as non-parties through the use of harmonized inventories.

One requirement would be that subnational entities within non-parties to the agreement meet MRV standards in order to maintain a robust market price as well as to ensure that their contributions to national targets are appropriately accounted for. Thus, a key starting point would be establishing common MRV and accounting methodologies to allow for harmonized inventories between the multiple levels of governance (Corfee-Morlot et al., 2009; Hoornweg, Sugar, & Gomez, 2011). There are already tools for inventories that allow for the break-down of carbon drivers and emissions (Lefèvre & Wemaere, 2009), which could be widely and cost-effectively adopted (Hoornweg et al., 2011). A CDM-type financing mechanism is unlikely to be useful as climate financing is targeted towards developing countries. Multilateral financing for non-party subnational actors in developed countries (the US) would be politically rejected. Allowing subnational actors in non-parties to access and use international carbon markets and build other market-based linkages is the only feasible form of a subnational financial link.

Since states drive and pioneer federal-level policy, a subnational market link which empower states and assists state-level action could create feedback which drive the adoption of federal-level climate legislation, as depicted in Figure 2. The first reinforcing feedback loop (R1) visually depicts how upwards policy diffusion and political pressure from states drives US multilateral engagement, which in turn leads to further investment and financing towards the subnational link as international actors see the positive effects. The second virtuous cycle (R2) illustrates that as state actors gain from access to markets, financial resources, and policy diffusion they become institutionally enmeshed and demand more from the link.

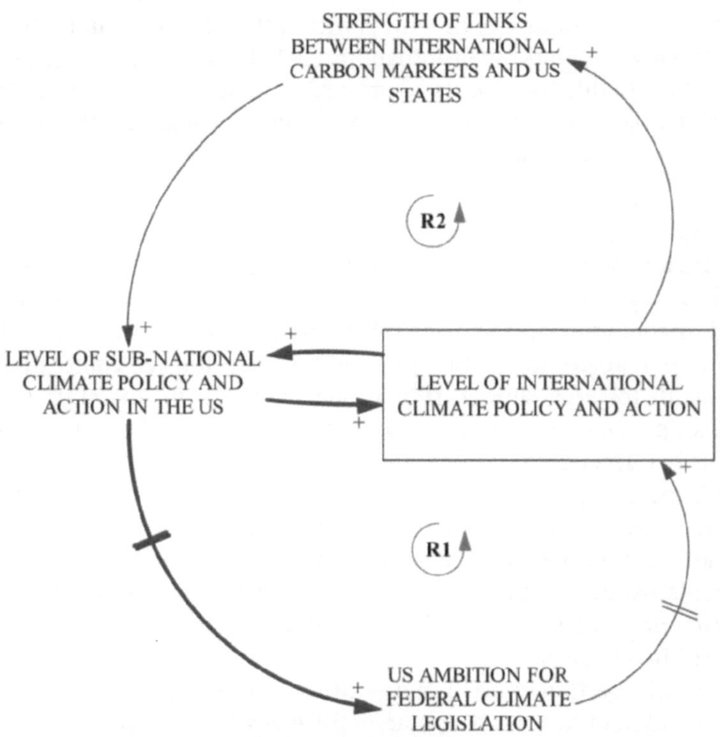

Figure 2. The subnational link and feedback to the federal level

5. Is Paris US-proof?

Given the existing options for addressing US non-party status is the current Paris Climate Agreement likely to be able to effectively cope if the US withdraws from the treaty or does not legally adopt it? Unfortunately for now the answer is no. None of the tools outlined in the previous section is mentioned in the Paris Agreement. As noted previously, the Paris Agreement is explicitly facilitative and non-punitive and provides no provisions relating to non-parties (nor does the UNFCCC).

While there are provisions for involving subnational actors under sections IV and V of the COP21 decision, these are weak and superficial. For example, paragraph 135 simply invites different non-state stakeholders to scale up existing actions and register these via the NAZCA platform. Article 6 of the agreement outlines mechanisms for cooperation, including market-based mechanisms. Under Article 6 market mechanisms are not directly mentioned, but instead referred to as 'internationally transferred mitigation outcomes'. Paragraph 3 stipulates that these internationally transferred mitigation outcomes will be used to meet nationally determined contributions (NDCs) under the agreement and 'shall be voluntary and authorized by participating Parties.' The direct reference to participating Parties and the linking of market mechanisms to achieving pledges under the agreement would preclude the involvement of non-parties and subnational actors within them. Paragraph 4(b) outlines that the aim of the sustainable development mechanism under the agreement is 'To incentivize and facilitate participation in the mitigation of greenhouse gas emissions by public and private entities authorized by a Party'. Article 6 currently only enables Parties with NDCs to participate in market-based and other mechanisms. Amendments to the agreement would be necessary to allow for the use of a subnational market link to a non-party US. Aside from Article 6, Article 28 specifies that withdrawal from the agreement can only take place 3 years after entry into force for that party. If the Paris Agreement enters into force just prior to a Republican president it could delay their intent to withdraw. But this would be short-lived, lasting less than one presidential term, and could be circumvented by a president simply withdrawing from the UNFCCC which would result in an automatic withdrawal from the Paris Agreement.

The Paris Climate Agreement cannot be considered US-proof: none of the tools to manage a non-party US have been enshrined into the text. The Paris Agreement is reliant on the will of the existing President and good intentions of future presidents. It should also be noted that the agreement only establishes a 'facilitative' and 'non-punitive' compliance mechanism under Article 15. If a future President was to abolish domestic executive measures such as the Clean Power Plan, it would likely lead to the US missing its proposed targets and damaging the legitimacy of the Paris Agreement. Even if a future Republican executive chooses to stay party to the Paris Agreement there is still no way of effectively handling intentional non-compliance. It is a scenario which could be just as damaging as a US withdrawal.

If the Paris Climate Agreement is to effectively deal with a non-party US then amendments to the agreement will be necessary. While the Paris Agreement can be amended through a three-quarters majority vote as per Article 22, such a move faces political obstacles. Parties have traditionally been unwilling to amend existing legal instruments for fear of 'reopening' agreements and allowing for a flood of new proposals. However, some amendments, such as the 2012 Doha amendments to the Kyoto Protocol, have been previously agreed to by consensus. Second, amendments are only binding upon those Parties which accept them. Non-universal adoption of amendments could lead to a legally confusing scenario. However, this was a common practice under the Montreal Protocol with subsequent amendments often not incurring universal ratification. While the reform of the Paris Climate Agreement to allow for mechanisms to address a non-party US faces challenges, it is feasible.

6. Architectures for managing a non-party US

The suite of tools for addressing US non-party status can be combined to craft different approaches. This section will explore two broad models: (a) the existing pledge and review system of the Paris Agreement amended to allow for subnational market link and (b) a carbon budget inclusive of non-party subnational actors with more stringent non-party measures.

6.1. Amending the Paris Agreement

The existing pledge and review system of the Paris Agreement could be amended to increase its resilience to a non-party US. As noted, facilitative non-party measures would likely be useless and punitive measures appear highly unfeasible. The only feasible and effective option would be to establish a market link to subnational actors in non-parties. This would require amendments to Article 6 and the development of further rules under future COP decisions. This could allow for subnational states in the US to join carbon markets and efforts outside the US. This could encourage bottom-up action in the US, both helping to heighten climate action domestically in the US and aiding subnational states like California to lobby and pressure the federal government. While such a market link could help to functionally involve a non-party US, it is debatable how much it could help maintain the legitimacy of a Paris Agreement without federal-level US participation.

6.2. An alternative to the Paris Agreement

An alternative to the US-vulnerable Paris Agreement is the scenario of 'Exclusive Critical Mass Governance' (ECMG). This is a model whereby a small but powerful group of major emitters (most likely the EU and China) go ahead and forge their own semi-global agreement without US participation. It is similar to my previously proposed model of 'Critical Mass Governance' which was intended to allow for US legal participation in an effective climate treaty (Kemp, 2015a). This proposal could be thought of as 'Inclusive Critical Mass Governance' in contrast to the ECMG model outlined here, as both attempt to craft effective climate architectures, but one does so with US legal participation and the other without. Both rely on a plurilateral logic that an agreement can start narrow-and-deep and broaden in membership and deepen in substance over time (Kemp, 2015a).

In this case, a coalition of parties could forge ahead and create an ambitious top-down framework based upon a carbon budget model. Fears over carbon leakage have little to no empirical basis (Branger & Quirion, 2014) and any potential leakage could be addressed through punitive non-party measures such as BTAs. The agreement would also utilize the subnational market link in order to allow for greater cost-effectiveness by maximizing mitigation opportunities while maintaining ties with the US. In that case, the participating subnational entities would need to have both requisite MRV standards as well as take targets in line with their fair share under the carbon budget. Periodic review via a framework for subnational actors could help ensure this and facilitate compliance. If the carbon budget was able to incorporate enough subnational actors from non-parties, it could develop a large degree of participation despite limitations in national actor numbers. Within this scenario the idea would be that the pressures of subnational action and trade restrictions would eventually push the US to ratify it. This approach has some resemblance to the 'climate club' proposal of Victor (2011) but differs in its use of a carbon budget, initial exclusion of the US, and use of non-party measures paired with a subnational financial link.

The implementation of ECMG in its purest form faces a number of potent political and technical obstacles. First and foremost, it would require strong leadership from China and/or the EU, neither of which currently appears willing to do so. China, in particular, has made it abundantly clear through official statements and submissions that it expects developed countries to take the lead in mitigation action. China, while capable, appears largely unwilling to take a leadership role before the US. Based on this, China would be unlikely to participate in any agreement that lacks US legal participation.

Yet the domestic actions of China often do not match its international rhetoric. The ambitious domestic policies of China have been driven by domestic co-benefits such as reducing air pollution, improving energy security, and gaining an economic advantage in the renewable energy industry (Teng & Jotzo, 2014). This string of strong domestic actions has intensified with China recently announcing the creation of national emissions trading scheme beginning in 2017 and the banning of new coal mines until 2019. Thus, it is not out of the question that China may take a strong leadership role based on perceived domestic benefits and choose to forge ahead without the US. The likelihood of this extending to approaches as drastic as trade measures appears highly unlikely.

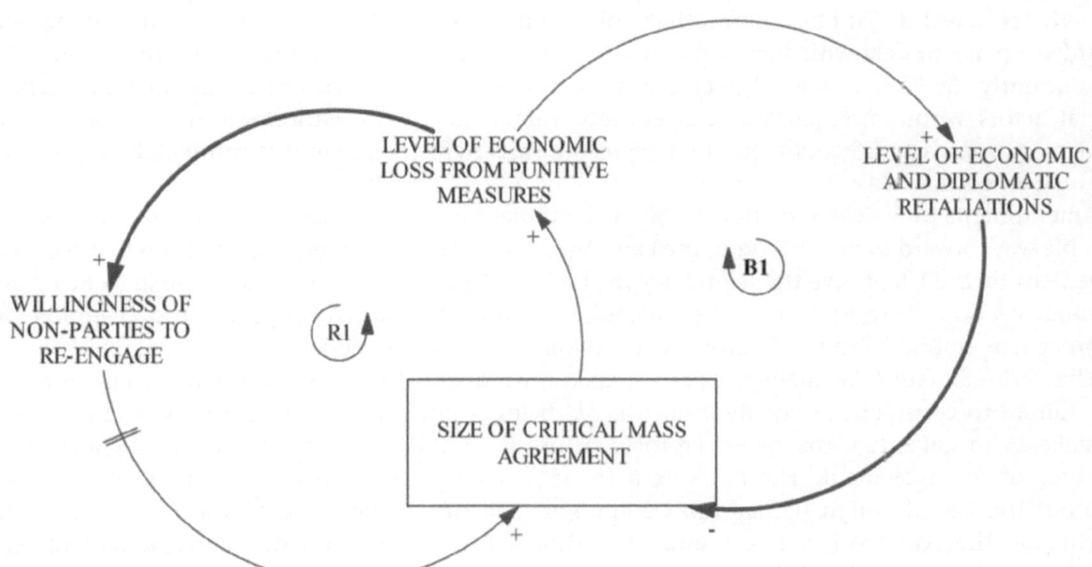

Figure 3. International feedback and US engagement

As shown in Figure 3, trade restrictions can encourage greater willingness to cooperate as the affected party wishes to alleviate economic losses. If the impacted party decides to reengage then a reinforcing feedback is created as the power of penalties increases with more countries joining the agreement (R1). However, trade measures would likely trigger diplomatic and economic retaliations. This tight balancing feedback (B1) deters such action, particularly by major developing countries such as China who have previously been unwilling to discuss trade restrictions under the UNFCCC (Aakre, 2014). Second, and most importantly, the Paris Agreement has already been established and the creation of a second international framework is now a forlorn hope. Third, the UNFCCC works by consensus and therefore decisions over entry into force requirements, non-party measures, and the broader architecture would need to be agreed to by all parties (Hovi, Skodvin, & Aakre, 2013). While there are legal opportunities for the UNFCCC to adopt voting, which could allow for such a plurilateral approach, these appear currently unfeasible (Kemp, 2015b). As Hovi et al. (2013) argue the only options would be for such a coalition and legal framework to be developed outside the auspices of the UNFCCC. While this approach seems unfeasible for now it could become the basis of a future deal between China and the EU outside of the UN. It could provide an effective way of marginalizing the US and addressing leakage concerns, whilst encouraging their longer-term reengagement. In any case, such ideal architectures are important to keep in mind as benchmarks for how an agreement could be designed to address a non-party US.

7. Conclusion

While the US is likely to ratify the Paris Agreement via a presidential-executive agreement, there is nonetheless a real threat of US non-ratification or withdrawal under a future president. Having safeguards in place to deal with such a problem is a wise, precautionary approach with few trade-offs. Yet currently the Paris Climate Agreement has no non-party measures, or ways of including subnational actors within non-parties. It is entirely vulnerable to US withdrawal and non-parties in general. A non-party US could lethally wound the Paris Agreement, yet it is completely unprepared for such an occurrence.

Amendments to Article 6 to allow for the involvement of subnational actors in non-parties is one feasible way forward to help US-proof the Paris Agreement. This is unlikely to be enough to drive reengagement by the US or save the legitimacy of the Paris Agreement. Truly effective management of a non-party US is likely to require the unfeasible strategy of punitive measures and a semi-global approach, as embodied in the Exclusive Critical Mass Governance model.

The Paris Climate Agreement has been shaped towards the preferences and circumstances of the US. Failure to consider the possibility of the US being a non-party in the future would be a fatal mistake. A mistake, like the Kyoto Protocol before it, could jeopardize the legitimacy and effectiveness of the agreement. The tools exist to help ensure that the Paris Agreement can operate without the US, although it may not be enough. The world may need to learn the lesson that relying on the good will of one president is short-sighted. Environmental agreements will only succeed in the long term if they take seriously the threat of non-parties, particularly if they are superpowers.

Acknowledgements

I would like to thank Janette Lindesay, Karen Hussey, Rob Dyball, Lea-Sofia Luzzi, Saskia McCulloch, and Tom Swann for their invaluable comments and edits.

Disclosure statement

No potential conflict of interest was reported by the author.

Notes

1. Concurrent Resolution, Senators Lee and Kelly, Senate Congressional Resolution to the 1st session of the 114th Congress, DAV15K66.
2. This equates to a cut of 12–19% below 1990 levels.
3. Unless it is the case that the Congress has abolished the executive agreement and the president remains enthusiastic to engage multilaterally.

References

Aakre, S. (2014). The political feasibility of potent enforcement in a post-Kyoto Climate Agreement. *International Environmental Agreements: Politics, Law and Economics*, 1–15. doi:10.1007/s10784-014-9238-5

van Asselt, H., & Brewer, T. (2010). Addressing competitiveness and leakage concerns in climate policy: An analysis of border adjustment measures in the US and the EU. *Energy Policy, 38*, 42–51. doi:10.1016/j.enpol.2009.08.061

Bang, G. (2010). Energy security and climate change concerns: Triggers for energy policy change in the United States? *Energy Policy, 38*(4), 1645–1653.

Bang, G., Froyn, C. B., Hovi, J., & Menz, F. C. (2007). The United States and international climate cooperation: International "pull" versus domestic "push". *Energy Policy, 35*(2), 1282–1291.

Bang, G., Hovi, J., & Sprinz, D. F. (2012). US presidents and the failure to ratify multilateral environmental agreements. *Climate Policy, 12*(6), 755–763.

Barker, T., Bashmakov, I., Bernstein, L., Bogner, J. E., Bosch, P. R., Dave, R., & Davidson, O. R. (2007). 11.7.2 Carbon leakage: Mitigation from a cross-sectoral perspective. In B. Metz, O. R. Davidson, P. R. Bosch, R. Dave, & L. A. Meyer (Eds.), *Climate change 2007: Mitigation. Contribution of working group III to the fourth assessment report of the intergovernmental panel on climate change* (pp. 76–81). Cambridge, UK: Cambridge University Press.

Betsill, M., Dubash, N. K., Paterson, M., van Asselt, H., Vihma, A., & Winkler, H. (2015). Building productive links between the UNFCCC and the broader global climate governance landscape1. *Global Environmental Politics*. doi:10.1162/GLEP_a_00294

Bodansky, D. (2015). *Legal options for US acceptance of a new climate change agreement*. Arlington, Virginia: Center for Climate and Energy Solutions.

Branger, F., & Quirion, P. (2014). Climate policy and the 'carbon haven' effect. *Wiley Interdisciplinary Reviews: Climate Change, 5*(1), 53–71.

Brunnee, J. (2008). Europe, the United States, and the global climate regime: All together now. *Journal of Land Use & Environmental Law, 24*, 1–44.

Burtraw, D., & Woerman, M. (2012). *US status on climate change mitigation* (Resources for the Future Discussion Paper No. 12–48).

Burtraw, D., & Woerman, M. (2013). Economic ideas for a complex climate policy regime. *Energy Economics, 40*(Supplementary 1), S24–S31. doi:10.1016/j.eneco.2013.09.021

Chan, S., & Pauw, P. (2014). *A global framework for climate action: Orchestrating non-state and subnational initiatives for more effective global climate governance.* Bonn: Deutsches Institut für Entwickslungpolitik.

Chang, H. (2010). International executive agreements on climate change. *Columbia Journal of Environmental Law, 35*(2), 337–371.

Corfee-Morlot, J., Kamal-Chaoui, L., Donovan, M., Cochran, I., Robert, A., & Teasdale, P.-J. (2009). *Cities, climate change and multilevel governance.* Paris: OECD.

EPA (Environmental Protection Agency). (2014). *Inventory of U.S. greenhouse gas emissions and sinks 1990–2012.*

Falkner, R. (2005). American hegemony and the global environment. *International Studies Review, 7*(4), 585–599.

Farber, D. A. (2014). Climate policy and the United States system of divided powers: Dealing with carbon leakage and regulatory linkage. *Transnational Environmental Law, 3*(01), 31–55. doi:10.1017/S2047102513000186

Grundig, F., & Ward, H. (2015). Structural group leadership and regime effectiveness. *Political Studies, 63*(1), 221–239.

Haites, E., Yamin, F., & Höhne, N. (2013). *Possible elements of a 2015 legal agreement on climate change.* Paris: IDDRI.

Hoornweg, D., Sugar, L., & Gomez, C. L. T. (2011). Cities and greenhouse gas emissions: Moving forward. *Environment and Urbanization, 23*(1), 207–227.

Hovi, J., Skodvin, T., & Aakre, S. (2013). Can climate change negotiations succeed? *Politics and Governance, 1*(2), 138–150.

Ismer, R., & Neuhoff, K. (2007). Border tax adjustment: A feasible way to support stringent emission trading. *European Journal of Law and Economics, 24,* 137–164. doi:10.1007/s10657-007-9032-8

Kemp, L. (2015a). Bypassing the 'ratification straitjacket': Reviewing US legal participation in a climate agreement. *Climate Policy.* doi:10.1080/14693062.2015.1061472

Kemp, L. (2015b). Framework for the future? Exploring the possibility of majority voting in the climate negotiations. *International Environmental Agreements: Politics, Law and Economics,* 1–23. doi:10.1007/s10784-015-9294-5

Krutz, G. S., & Peake, J. S. (2009). *Treaty politics and the rise of executive agreements: International commitments in a system of shared powers.* Ann Arbor, MI: University of Michigan Press.

Kulovesi, K. (2014). Real or imagined controversies – a climate law perspective on the growing links between the international trade and climate change regimes. *Trade, Law and Development, 6,* 55–92.

Lachapelle, E., & Paterson, M. (2013). Drivers of national climate policy. *Climate Policy, 13*(5), 547–571.

LefÈvre, B. (2012). Incorporating cities into the post-2012 climate change agreements. *Environment and Urbanization, 24*(2), 575–595.

Lefèvre, B., & Wemaere, M. (2009). *Post-2012 climate change agreement-fitting commitments by cities: Political, economic, technical and legal aspects.*

Neslen, A. (2016, February). Donald Trump warned against scrapping Paris climate deal. *The Guardian.*

Patrick, S. (2002). Multilateralism and its discontents: The causes and consequences of US ambivalence. In S. Patrick & S. Forman (Eds.), *Multilateralism and US foreign policy: Ambivalent engagement* (pp. 1–44). Boulder, CO: Lynne Rienner.

Pauwelyn, J. (2013). Carbon leakage measures and border tax adjustments under WTO law. In G. Van Calster & D. Prévost (Eds.), *Research handbook on environment, health and the WTO* (pp. 448–506). Cheltenham, UK: Edward Elgar.

Peake, J. S., Krutz, G. S., & Hughes, T. (2012). President Obama, the senate, and the polarized politics of treaty making*. *Social Science Quarterly, 93*(5), 1295–1315. doi:10.1111/j.1540-6237.2012.00913.x

Purvis, N., & Stevenson, A. (2010). *Rethinking climate diplomacy.* Washington, DC: German Marshall Fund.

Rabe, B. (2008). States on steroids: The intergovernmental odyssey of American climate policy. *Review of Policy Research, 25*(2), 105–128.

Rabe, B. (2011). Contested federalism and American climate policy. *Publius: The Journal of Federalism, 41*(3), 494–521.

Rabe, B., Roman, M., & Dobelis, A. (2005). State competition as a source driving climate change mitigation. *NYU Environmental Law Journal, 14*, 1–53.

Schreurs, M. A. (2008). From the bottom up: Local and subnational climate change politics. *The Journal of Environment & Development, 17*(4), 343–355.

Schreurs, M. A., Selin, H., & VanDeveer, S. D. (2009). Expanding transatlantic relations: Implications for environment and energy politics. In M. A. Schreurs, H. Selin, & S. D. Vandeveer (Eds.), *Transatlantic environment and energy politics: Comparative and international perspectives* (pp. 1–21). Farnham: Ashgate.

Selin, H., & VanDeveer, S. D. (2011). US climate change politics and policymaking. *Wiley Interdisciplinary Reviews: Climate Change, 2*(1), 121–127.

Teng, F., & Jotzo, F. (2014). Reaping the economic benefits of decarbonization for China. *China & World Economy, 22*(5), 37–54.

Tjernshaugen, A. (2005). *United States participation in future climate agreements: An assessment* (Policy Note). CICERO-Senter for klimaforskning.

Veel, P.-E. (2009). Carbon tariffs and the WTO: An evaluation of feasible policies. *Journal of International Economic Law, 12*(3), 749–800.

Victor, D. (2011). *Global warming gridlock: Creating more effective strategies for protecting the planet*. Cambridge: Cambridge University Press.

Ye, J. (2014). *Comparison of carbon pricing proposals in the 113th Congress*. Arlington, VA: Centre for Climate and Energy Solutions.

Zhang, Z. (2009). Multilateral trade measures in a post-2012 climate change regime? What can be taken from the Montreal Protocol and the WTO? *Energy Policy, 37*(12), 5105–5112.

Global trade and promotion of cleantech industry: a post-Paris agenda

JOHN A. MATHEWS

The Paris Agreement adopted in December 2015 puts the world on a new footing for global solutions to the issue of mitigation of climate change. In the formulation of climate policy, the issues of trade in cleantech goods, and the promotion of cleantech industries through industrial policy measures such as local content requirements, now loom large. The case for reconciliation between the United Nations Framework Convention on Climate Change (UNFCCC) and World Trade Organization (WTO) is now becoming stronger, and calls for specific steps to enable trade to accommodate green industry promotion strategies. Specifically, it is proposed that the next Conference of the Parties under the UNFCCC, to be staged in Morocco late in 2016, adopt a resolution calling on the WTO to recognize the authority of the UNFCCC in recognizing climate change mitigation as a 'public good', whose pursuit might allow for temporary use of industrial development policies such as local content requirements, if countries wished to follow such a strategy.

Policy relevance

Countries' trade policies are currently formulated without reference to climate change, while climate change strategies are formulated without reference to principles of fair trade. The article proposes means through which the two regimes, governing trade and climate change, and the two principal bodies involved, the WTO and UNFCCC, can be brought to a state of mutual recognition and (prospective) harmony. The implications for climate policy are traced.

1. Introduction

The Paris Agreement reached in December 2015 is widely viewed as representing a fresh start for the mitigation of climate change through global measures. While falling short of setting a global price on carbon, nevertheless the Climate Agreement reached in Paris registers the world's first truly global commitment to reduce collective carbon emissions.[1] As such, it opens the way to more detailed discussions as to how countries might implement the Intended Nationally Determined Contributions (INDCs) to which they have agreed. Central to this process will be the role played by the promotion of cleantech industries, particularly renewable energy industries, and the interaction between national promotion measures such as the imposition of local content requirements (LCRs) and the global trade regime.

A rash of disputes over green energy promotion, both brought before the World Trade Organization (WTO) and by countries imposing unilateral 'trade remedies' on each other, threaten to subdue countries' enthusiasm for building green industries. These disputes have come to a head in the past few years, leading scholars to label them as the 'next generation' of trade and environment conflicts and viewing them as imposing a serious risk of reversing whatever gains might be made in the forums established within the terms of the United Nations Framework Convention on Climate Change (UNFCCC) (Bigdeli, 2014; Wu & Salzman, 2014).

Indeed, there is a very real danger that actions brought against countries in the name of a narrow interpretation of WTO rules might jeopardize the gains won at Paris. What is won within the arena of the UNFCCC might be lost again within the arena of the WTO. How is this highly undesirable outcome to be avoided?

The rash of trade disputes trace their origin to the rapid rise of green industrial policies being pursued not only by new industrial powers such as China, India, Brazil and South Africa, but also by state- and provincial-level governments in the EU, Canada and the US. While legal scholars have advanced various ways of dealing with these disputes and their serious ramifications for efforts to advance decarbonization, these suggestions have mostly involved reforms to WTO law and its various recognized exceptions, as well as to national-level trade remedy processes such as the imposition of Countervailing Duties (CVDs).

What is missing in the various proposals advanced so far is an element of supranational endorsement of any particular country's imposition of promotional measures such as LCRs. Such a policy goal might be pursued in the name of a global public good in the form of mitigation of climate change through the promotion of green energy and cleantech industries. The goal would be given so much added credibility if it could appeal to the newly adopted Paris Agreement as a clearly stated collective aspiration of humanity.

This article makes a clear and straightforward proposal that there be mutual recognition of each other by the WTO and UNFCCC, which would be of benefit to both parties. The UNFCCC's Paris Agreement would thereby gain the 'teeth' needed to allow countries to translate general decarbonization commitments into green industries promotion. And the WTO, by recognizing the authority of the UNFCCC in designating climate change as a matter of global public interest, would have the means of allowing for the promotion of green industries without undermining its whole rationale and avoiding being seen as obstructing the development of solutions to humankind's greatest challenge. This article provides consideration and analysis of such a way forward beyond the Paris Agreement.

2. 'Next generation' trade disputes over green products and imposition of LCRs

The imposition of LCRs as a means of promoting domestic industries has a long history. Known variously as 'offsets' or non-tariff trade barriers, LCRs have typically been attached as conditions to public procurement contracts and have usually been imposed within a strict time frame. They are viewed favourably by development economists who see a useful role for the state in driving industrial development (see e.g. Amsden, 1989; Wade, 1990), but unfavourably by economists who view them as interfering with the claimed benefits of free trade (Hufbauer, Schott, & Cimino, 2013). Now in the case of green industries they are proving their worth again, but are being curbed by actions brought to the WTO by the US, EU and Japan.

The use of LCRs in building green industries by China, India, etc., dates back to China's active development of its wind power industry in the 1990s. In the country's 9th Five-Year Plan (covering the years 1996–2001), there were targets for wind power development and local content targets that came to be formally implemented in the Wind Power Concession Programme launched in 2003 by the National Development and Reform Commission (NDRC). This programme targeted larger scale projects of between 100 and 200 MW and called for concession projects to be linked to 40% (rising to 70%) local content, with the specific goal of building a local green industry. Under international protest, the programme was discontinued in 2009 (revoked by NDRC Notice 2991 in November 2009) – but already the programme had had its effect, and was being superseded by financing arrangements that likewise required local wind power projects to meet certain specified grid-connection targets in order for funding to be approved. There is widespread agreement that this was a highly successful programme and one that is largely responsible for China's wind power generation and wind turbine manufacturing sector becoming world number one in the 2000s.[2] It did not result in a formal WTO dispute because China discontinued the programme after judging it to be a success.

The 2000s saw many more LCR provisions being implemented, notably by EU countries such as Spain and by US states including California and Canadian provinces such as Ontario and Quebec. They have provoked numerous counteractions taken within the trade arena. Formal action was in fact taken by Japan against Ontario's Feed-in-Tariff (FiT) provisions coupled to LCRs, which, despite their success in helping to build an effective renewable energy system in the province and reducing carbon emissions, were made the subject of complaint by Japan and deemed WTO-incompatible by an Appellate Body hearing in 2013.[3] There have been further actions taken against China's promotion of both its wind turbine industry, through LCRs, and its solar photovoltaic (PV) industry through local subsidies and tax breaks (in separate actions brought by the US and the EU); and against India's National Solar Mission (NSM), which was designed to bring India up from being a laggard to being a leader in transitioning to a clean energy future (an action brought by the US). Meanwhile, China has itself brought cases at the WTO against US state-level LCRs and local subsidies, while the US has responded to domestic pressures and imposed Anti-dumping (AD) and CVDs on two occasions against Chinese PV imports, while China has retaliated by imposing CVDs on US exports of such products as PV cell-making equipment and silicon exports.

Brazil has utilized a smart combination of strategies to enable it to catch up in renewables and build its own renewable power industries. These strategies include guaranteed power purchase agreements (PPAs) and power-contract auctions to boost the market for renewables, as well as the use of indirect local sourcing requirements not imposed through the trading system (where they would attract attention from trading partners) but instead indirectly through the financing mechanism, operated by the Brazilian National Development Bank (BNDES).

The case of India may be taken as exemplary, both because it stands as a clear model for other countries and because its details have been reviewed by a panel of the WTO. The LCR provisions of the NSM are quite explicit – and have provided a template also for the proposed National Wind Program (NWM). The original NSM was launched by the Singh administration with comprehensive national development goals being made clear (Johnson, 2016; Kent & Jha, 2014). An initial total of

10 GW was expected to be installed over five years. This aspect has been drastically upgraded to reach a new 100 GW target by 2019. India has learned from China the power of LCRs as a tool for domestic industry development.[4]

India's NSM was challenged at the WTO by the US, leading to the convening of an Appellate Body that reported to the parties in August 2015 and with a final report being issued publicly in February 2016 (WTO, 2016). This report finds against India, and rebuts each of India's defences of its LCRs under Article XX of the General Agreement on tariffs and Trade (GATT) and under Article III.4 of the GATT, where India argued that it was required under international law to provide green electricity and that its LCRs as part of the NSM were designed to enable India to comply with this requirement. The report acknowledges India's right to seek a way forward that enhances its energy security and reduces its carbon emissions, but then dismisses the LCRs as unlikely to achieve their green goal.

Now, in the wake of the UNFCCC Paris Agreement (once it is ratified), there is a much stronger justification for countries to engage in green industry-building activities – if they wish to do so. Of course many countries would be content to import green technologies from another country that can produce them cost-effectively. But many countries would see the building of green industries as a convenient means of achieving their commitments while contributing to the global 'green shift' that is under way. For such countries, a global statement that the WTO recognizes the UNFCCC as a body with the authority to stipulate that green industry-building is an activity that meets the test of being a 'public good' – and that therefore GATT Article 20 exemptions would apply if appealed to – would be helpful. This should not be viewed as 'suspending' free trade rules, but as clarifying the circumstances in which countries could claim exemption under GATT Article 20 rules for a designated period and facilitate the choice by countries such as India in their bid to develop clean industry capabilities.

It is clear that most of the major industrializing powers are now embarking on serious build-up of their renewables sectors, and that they are doing so through judicious use of green industrial strategies. How then are these developments to be made compatible with the world trading system?

3. Potential green reforms to the world's trade system

In the absence of reforms to the general trade system, specific products have been targeted by groups of countries and adopted as tariff-reducing agreements by the WTO. There is under way another such initiative aimed at creating an Environmental Goods Agreement (EGA) which would reduce tariffs on a range of goods deemed 'environmental' or green – such as renewable energy devices. A 'group of 14' countries have been hammering out agreement on such an EGA, for adoption by the WTO, and received a boost following the meeting of the G20 at Hangzhou in China in September 2016, when the goal of producing a final text of the EGA by the end of 2016 was spelt out.[5] The proposal builds on the great success of the prior WTO agreement on IT goods, which has underpinned these goods' vast market expansion since the 1980s.[6] The current Geneva negotiations on a comparable WTO agreement on free trade in green goods build on the Asia Pacific Economic Cooperation (APEC) agreement reached in Vladivostok in September 2012, which required APEC member countries to reduce tariffs on green goods down to less than 5% by 2015.[7]

Now genuine free trade in green goods (including intermediate parts) is clearly superior in its market-expanding effects to targeted instruments such as LCRs. So countries like India would be able to benefit from joining such a WTO-endorsed 'free trade in green goods' agreement, and correspondingly repealing their LCR measures on green goods. But in the meantime, industrial measures such as LCRs will help countries like India to build the green industries which will produce the goods that can eventually be traded free of tariffs and other restrictions.

Countries that seriously tackle the challenge of decarbonizing their energy systems need to employ some form of green industrial policy. They utilize one or more of the various industrial strategies involving subsidies, low-interest loans, tax breaks and above all FiTs and LCRs which have proven themselves to be powerful instruments, when employed judiciously, to build new cleantech value chains and producers of cleantech products. It is not a question of allowing newly industrializing countries like China and India to build renewable energy industries that will then destroy such industries in already developed countries (Chufart-Finsterwald, 2014; Frey, 2016). Rather, it is a question of promoting the diffusion of cleantech industries and encouraging the planting of such industries in countries around the world, subject to WTO oversight to ensure that competition remains reasonable and does not become predatory.[8]

The few scholars who examine these trends in world trade and the disputes arising have some reforms to offer, both to the operation of the WTO system itself and to the imposition of trade remedies by countries individually (as the US has done in its dispute with China over solar PV imports) (Voituriez & Wang, 2015). In the 'Greenprint' offered in realistic fashion by Indian-US scholars, Mattoo and Subramanian (2012, 2013) frame the problem (as done here) as requiring a fresh start for climate negotiations based on the recognition of the now dominant role played by the new industrial giants China, India and Brazil, and recognizing that a change in WTO rules is needed and justified to allow countries to take the required steps to build green industries. They sensibly argue that production subsidies for specific green products and technologies – paid under such provisions as FiTs and direct support for renewable energy industries – be deemed permissible for a specified period. Export subsidies should not be prohibited, but depending on the circumstances should be merely 'actionable' (meaning they can be challenged if a case can be made). Without specifically mentioning LCRs, Mattoo and Subramanian paradoxically recommend no change for existing trade rules covering 'subsidies contingent upon the use of domestic over imported goods'. This would allow the current spate of trade disputes over LCRs to continue unabated, and possibly exacerbate them. So the proposal advocated in this article departs from that of Mattoo and Subramanian in this particular respect.

What is really needed is a process that cuts through various vested interests and allows the principal issues before the two global bodies concerned, the UNFCCC and the WTO, to be directly addressed. That is precisely the kind of proposal advanced in this article.

4. Integrating the world's trade and climate regimes

The few examples of scholarly suggestions for integrating the role of the WTO with the climate change regime focus on expanding the role of exceptions to cover the public interest 'good' of climate change mitigation. Howse (2010), for example, argues that the existing Article XX of the GATT could be expanded easily enough to cover 'climate change mitigation' so that countries introducing measures to support green industries would seek exemption on these grounds under the GATT Art XX that lists measures 'necessary to protect human, animal or plant life or health' as allowable exceptions.[9] This would make it harder to bring an action against them – but not impossible if the party is determined enough. The reason for the fallibility of such a defence is that the country introducing the measure is taking unilateral action and offering no more than its own assurance that the measure is protected by Article XX. Something broader is clearly called for to stem the potential 'rash' of green-related disputes.

One can agree with Howse (2010) and with Mattoo and Subramanian (2013) that that 'something broader' should be the imprimatur of the UNFCCC itself stipulating that certain measures taken in order to green industry are in the global public interest and deserve to be protected from action under WTO rules.[10] A resolution adopted by a future COP (starting with the COP 22 to be staged in Morocco in late 2016) could list such measures in an open and transparent fashion. It would be open to the WTO as a global body to state that it is prepared to recognize such a UNFCCC-based schedule as specifying measures that are taken in the public interest and are to be exempted from the usual rules of competition for a specified period (time, place and coverage).

Therefore, one practicable way forward would be for the next COP (COP 22 to be staged in late 2016 in Morocco) to adopt a resolution recognizing a group of green products and processes as being eligible for exemption from trade law for a set period (say five years), a period which would enable countries that wish to do so to take advantage of a broad but limited exemption from trade rules in order to build their green industries. Rather than leave individual countries on their own to argue their case before the WTO, such a resolution would lend the authority of the UNFCCC to such matters, cutting through the complications raised by WTO procedures and establishing once and for all that climate change mitigation is the ultimate 'public interest' that needs to balance the stringent application of trade provisions with exemption for certain designated products and periods from the rules of full competition.[11]

The power of this proposal lies in the point that it promotes the diffusion of green industries through allowed imposition of green industrial policy; it does not undermine WTO principles of free and fair trade, but simply allows the WTO to (implicitly) recognize the UNFCCC as providing the authority needed to specify products or processes as being green, and give countries that wish to do so a window of protection to allow them to build the needed green industries. It would take the provisions of the newly adopted Paris Agreement beyond merely specifying reductions in carbon emissions towards engaging with the means that countries need to employ to embark on serious decarbonization of their energy systems.

5. Concluding remarks

The proposed reconciliation between the WTO and the UNFCCC would drive investment, trade and industry towards low-carbon and zero-carbon activities – seriously and irreversibly. It would finally unleash the forces of capitalist competition to drive technological and production innovation in favour of reducing carbon emissions, as called for by the WTO itself, the World Bank and the United Nations Conference on Trade and Development (UNCTAD) (Charnovitz, 2014; UNCTAD, 2014; WTO, 2011). The proposal calls for mutual recognition on the part of the UNFCCC and WTO with a view to generalizing and consolidating the grounds for GATT Article 20 exemption in the name of providing a 'public good' – where the public good would encompass climate change mitigation as a matter affecting humanity as a whole. This would have the effect of taking away the onus on individual countries to establish such an equation between climate change mitigation and the provision of a public good, if they elect to take the option of building their own green industries as a means of meeting their new obligations under the Paris Agreement. It would also recognize that some countries would elect to import green technologies as a means of meeting their commitments and would not need to appeal to exemption on grounds of the broadened notion of 'public good' as encompassing climate change mitigation. Either way the proposed accommodation between the WTO and UNFCCC would clarify countries' position with regard to trade matters and the benefits that can flow from an expansion of trade in green goods. This would surely provide a fitting way forward beyond the Paris Agreement.

Acknowledgements

I wish to acknowledge the very helpful comments made by the Editor in Chief at the proofs stage of the article which facilitated the argument being presented with greater clarity.

Notes

1. The Paris Agreement was adopted by the 21st Conference of the Parties (COP) meeting in Paris in December 2015, under the auspices of the UNFCCC. For background, see the Editorial in *Nature Climate Change* in December 2015 (Plotting a course from Paris, *Nature Climate Change*, *5*, 1021 (Dec 2015), as well as the text of the Agreement, available at the UNFCCC webpage: http://www.cop21.gouv.fr/wp-content/uploads/2015/12/l09r01.pdf.
2. On China's build-up of renewable energy industries, see Lewis (2014) or Mathews and Tan (2015), and on the sources of trade friction with other countries in the PV sector, see Dai and Xue (2015).
3. For the text of the WTO Appeal decision, see WTO (2014), and for comment, see Charnovitz and Fischer (2015).
4. On India's strategies including the use of LCRs, see Johnson (2016) and Chaudhary, Krishna, and Sagar (2015). For the report of the Appellate Body on the dispute involving India's LCRs as part of its NSM, see WTO (2016). It is worth noting that India has long used LCRs in other industries such as its automobile industry until discontinuing them after losing an appeal at the WTO. The LCRs framed more recently for the solar PV industry can be said to have embodied lessons learned in these earlier disputes – but still encountered resistance at the WTO.
5. The proposed Environmental Goods Agreement to be ratified by the WTO has been followed closely by the International Centre for Trade and Sustainable Development (ICTSD). See commentary on the G20 meeting statement and the initiative by the chair of the 'Group of 14' meeting in Geneva at: http://www.ictsd.org/bridges-news/biores/news/environmental-goods-agreement-negotiators-agree-roadmap-for-conclusion.
6. The process started as the IT Agreement reached by 29 participants at the Singapore Ministerial meeting in 1996; see the WTO record of this agreement at: https://www.wto.org/english/tratop_e/inftec_e/inftec_e.htm.
7. The green goods were identified as part of a schedule: http://apec.org/Home/Meeting-Papers/Leaders-Declarations/2012/2012_aelm/2012_aelm_annexC.aspx.
8. For overviews of the issues, see Rubini (2012); Low, Marceau, and Reinaud (2012); Cosbey and Mavroidis (2014); Kuntze and Moerenhout (2014); Wu and Salzman (2014); Meyer (2015); Leal-Arcas and Filis (2015) and Weiss (2016). For a recent discussion in light of the Paris Agreement of December 2015, see Hawkins (2016).
9. See the text of the General Agreement on tariffs and Trade (1947), GATT Art XX, at: https://www.wto.org/english/docs_e/legal_e/gatt47_02_e.htm#articleXX
10. Existing environmental measures not involving climate matters would remain unchanged.

11. See Esty (2016) for a recent comprehensive discussion of ways open to the WTO to expand its consideration of sustainability issues such as climate change mitigation.

Disclosure statement

No potential conflict of interest was reported by the author.

References

Amsden, A. (1989). *Asia's next giant: South Korea and late industrialization*. New York: Oxford University Press.

Bigdeli, S. Z. (2014). Clash of rationalities: Revisiting the trade and environment debate in light of WTO disputes over Green industrial policy. *Trade, Law & Development*, *6*(1), 177–209.

Charnovitz, S. (2014). *Green subsidies and the WTO* (policy research working paper #7060). Washington, DC: World Bank.

Charnovitz, S., & Fischer, C. (2015). Canada – renewable energy: Implications for WTO law on green and not-so-green subsidies. *World Trade Review*, *14*(2), 177–210.

Chaudhary, A., Krishna, C., & Sagar, A. (2015). Policy making for renewable energy in India: Lessons from wind and solar power sectors. *Climate Policy*, *15*(1), 58–87.

Chufart-Finsterwald, S. (2014). Environmental technology transfer and dissemination under the UNFCCC: Achievements and new perspectives. *Environmental Claims Journal*, *26*(3), 238–260.

Cosbey, A., & Mavroidis, P. C. (2014). A turquoise mess: Green subsidies, blue industrial policy and renewable energy: The case for redrafting the subsidies agreement of the WTO. *Journal of International Economic Law*, *17*, 11–47.

Dai, Y., & Xue, L. (2015). China's policy initiatives for the development of wind energy technology. *Climate Policy*, *15*(1), 30–57.

Esty, D. C. (2016). A proposed G20 Sustainability Package. In ICTSD, 2016. *G20 Hangzhou Summit 2016: Proposals for trade, investment, and sustainable development outcomes*. Geneva: International Centre for Trade and Sustainable development.

Frey, C. (2016). Tackling climate change through the elimination of trade barriers for low-carbon goods: Multilateral, plurilateral and regional approaches. In V. Mauerhofer (Ed.), *Legal aspects of sustainable development*. Springer International. doi:10.1007/978-3-319-26021-1_22

Hawkins, S. (2016, June). *Use trade policy to achieve the benefits of low emission development. Low Emission Development Strategies (LEDS)*. Geneva: International Centre for Trade and Sustainable Development.

Howse, R. (2010). *Climate mitigation subsidies and the WTO legal framework: A policy analysis*. Winnipeg: International Institute for Sustainable Development. Retrieved from http://www.cop21.gouv.fr/wp-content/uploads/2015/12/l09r01.pdf

Hufbauer, G., Schott, J., & Cimino, C. (2013). *Local content requirements: Report on a global problem*. Washington, DC: Peterson Institute for International Economics. Retrieved from http://biblioteca.fundacionicbc.edu.ar/images/b/b7/Hufbauer20131007ppt_local_content_proteccionism.pdf

Johnson, O. (2016). Promoting green industrial development through local content requirements: India's national solar mission. *Climate Policy*, *16*(2), 178–195.

Kent, A., & Jha, V. (2014). Keeping up with the changing climate: The WTO's evolutive approach in response to the trade and climate conundrum. *The Journal of World Investment & Trade*, *15*, 245–271.

Kuntze, J.-C., & Moerenhout, T. (2014). Are feed-in tariff schemes with local content requirements consistent with WTO law? In F. Baetens & J. Caiado (Eds.), *Frontiers of international economic Law: Legal tools to confront interdisciplinary challenges* (pp. 151–180). Nijhoff: Brill (Society of International Economic Law). Retrieved from http://booksandjournals.brillonline.com/content/books/b9789004270190_009

Leal-Arcas, R., & Filis, A. (2015). Renewable energy disputes in the World Trade Organization. *Oil, Gas & Energy Law Intelligence, 13*(3), 1–51.

Lewis, J. I. (2014). The rise of renewable energy protectionism: Emerging trade conflicts and implications for low carbon development. *Global Environmental Politics, 14*(4), 10–35.

Low, P., Marceau, G., & Reinaud, J. (2012). The interface between the trade and climate regimes: Scoping the issues. *Journal of World Trade, 46*, 485–544.

Mathews, J. A., & Tan, H. (2015). *China's renewable energy revolution*. London: Palgrave-Macmillan.

Mattoo, A., & Subramanian, A. (2012). *Greenprint: A New approach to cooperation on climate change*. Washington, DC: Center for Global Development.

Mattoo, A., & Subramanian, A. (2013). *Four changes to trade rules to facilitate climate change action* (CGD Policy paper 021). Washington, DC: Center for Global Development.

Meyer, T. (2015). *How local discrimination can promote global public goods*. Law School, Cornell University. Retrieved from http://www.lawschool.cornell.edu/cornell-IL-IR/upload/LocalDiscrimination-10-11-2015-clean-3.pdf

Rubini, L. (2012). Ain't wastin' time no more: Subsidies for renewable energy, the SCM agreement, policy space, and law reform. *Journal of International Economic Law, 15*, 525–579.

UNCTAD. (2014). *Trade remedies: Targeting the renewable energy sector*. Geneva: United Nations Conference on Trade and Development. Retrieved from http://unctad.org/en/PublicationsLibrary/ditcted2014d3_en.pdf

Voituriez, T., & Wang, X. (2015). Real challenges behind the EU-China PV trade dispute settlement. *Climate Policy, 15*(5), 670–677.

Wade, R. (1990). *Governing the market: Economic theory and the role of government in East Asian industrialization*. Princeton, NJ: Princeton University Press.

Weiss, E. B. (2016). Integrating environment and trade. *Journal of International Economic Law, 19*, 367–369.

World Trade Organization. (2011). *Harnessing trade for sustainable development and a green economy*. Geneva: Author.

World Trade Organization. (2014). *Canada: Measures relating to the feed-in tariff program, implementation notified by respondent, 15 June 2014*. Geneva: Author. Retrieved from https://www.wto.org/english/tratop_e/dispu_e/cases_e/ds426_e.htm

World Trade Organization. (2016, February 24). *India: Certain measures relating to solar cells and solar modules*. Report of the Panel. WT/DS456/R.

Wu, M., & Salzman, J. (2014). The next generation of trade and environment conflicts: The rise of green industrial policy. *Northwestern University Law Review, 108*(2), 401–474.

Index

Note: Figures are indicated by *italics*. Tables are indicated by **bold**. Endnotes are indicated by the page number followed by 'n' and the endnote number e.g., 20n1 refers to endnote 1 on page 20.